ADVANCES IN GENETICS RESEARCH

VOLUME 12

ADVANCES IN GENETICS RESEARCH

Additional books in this series can be found on Nova's website
under the Series tab.

Additional e-books in this series can be found on Nova's website
under the e-book tab.

ADVANCES IN GENETICS RESEARCH

VOLUME 12

KEVIN V. URBANO
EDITOR

New York

For permission to use material from this book please contact us:
Telephone 631-231-7269; Fax 631-231-8175
Web Site: http://www.novapublishers.com

Additional color graphics may be available in the e-book version of this book.

Library of Congress Cataloging-in-Publication Data

ISSN: 2159-1563

ISBN: 978-1-63321-351-7

Published by Nova Science Publishers, Inc. † New York

CONTENTS

PREFACE

"Advances in Genetics Research" presents original research results on the leading edge of genetics discovery. Each article has been carefully selected in an attempt to present substantial research results across a broad spectrum. In this continuing series compilation, the authors present and discuss varied topical data such as the "s" allele of the serotonin transporter and its lack of association with major depression in veterans; effects of regular versus special school placement in students with Down Syndrome; the biology of telomeres in multipotential stromal cells; genetic diversity of human and animal papillomaviruses; telomere length in lymphoid malignancies; telomerase and telomere maintenance in the blight fungus ustilago maydis; roles of mirnas in cancers; and genetic programming for music.

Chapter 1 – The results of some studies suggest that the serotonin transporter-linked polymorphic region (5-HTTLPR) short (S) allele, relative to the long (L) allele, is associated with risk for Major Depressive Disorder (MDD), and thus serves as a biomarker for MDD, while results from other studies do not support that conclusion. Persons with an S allele demonstrate a 2- to 2.5 fold decrease in serotonin transcription rate compared to the L-allele, which may increase their risk for MDD. Differences in study populations may help explain the differences in findings between those meta-analyses. To date, there have been no published reports which have addressed the possible association between the S allele and MDD among military veterans. This manuscript describes a first study to assess the possible association of the S allele with MDD among a study population of veterans in treatment for a substance use disorder. The authors hypothesized that the S allele would be associated with MDD in their study sample. Subjects signing informed consent were 101 Veterans recruited from VA behavioral health and substance use treatment clinics in the VA Pittsburgh Healthcare System, and 91 of those subjects were genotyped for 5-HTTLPR polymorphisms. The study sample from whom genetic material was collected included 82 males and 9 females, of whom 53 were white, 38 were black, and one was "other". Fifty-four members of the study sample (59%) met DSM-IV criteria for an MDD on the SCID. Forty-five of the subjects demonstrated one or two S alleles, while 46 did not do so. The presence of the S allele of the serotonin transporter was not found to be significantly associated with the diagnosis of major depressive disorder in the authors' sample (Chi-square=0.1.63, df=1, p=0.199). That finding, in combination with other recent negative findings from other researchers involving non-veterans, raises questions regarding the clinical utility of utilizing genetics tests involving the assessment of the alleles of the serotonin transporter as a possible biomarker for MDD.

Chapter 2 – *Background* Since the 1980s, more and more children with Down syndrome are in regular education. Parents and schools expect social and cognitive advantages.

Method This systematic review of studies on the effects of school placement of students with Down syndrome, with special reference to self-help skills, language, academics and social functioning, is based on the following criteria. Firstly, studies were published in the period 1970-2010. Secondly, any study with a direct comparison between placements in developmental or social outcomes was included. However, studies with a very small sample size (n<3) were excluded. Thirdly, non-comparative studies were included if in the study the acceptance of regularly placed children with Down syndrome by their own classmates was evaluated. Single case studies were excluded. Finally, studies were published in English, Dutch, German, French, Italian, Portuguese, Spanish, Norwegian, Swedish, Danish or Greek. To detect studies, comprehensive sources such as Picarta, Google, Medline, ERIC and Science Direct were used. In addition, a request for relevant research reports was sent out to all member organisations of the European Down Syndrome Organisation (EDSA).

Results The literature search yielded 133 potentially relevant studies, of which 53 met the inclusion criteria. In 26 studies, regularly and specially placed children with Down syndrome were compared and in another 27 studies, the acceptance of regularly placed children with Down syndrome by their classmates was evaluated. In nine of the 26 comparative studies, no attempt was made to disentangle the effect of selective placement from the effect of differences in stimulation between settings. However, in fifteen studies researchers did statistically correct for the effect of important other child and/or family variables that could have an impact on development. Furthermore, four studies can be considered to be natural experiments in which school placement was not determined by child characteristics but by geographical area and/or generation. Results show that regular classroom placement yields a better development of language and academic skills, even after the effect of selective placement has been taken into account. As regards self-help skills, under the same condition, there seem to be no differences between both types of school. Social functioning shows a mixed image. For social network, behaviour, and self-competence, no differences at all or small positive differences for regularly placed children were found. However, most studies also highlight that mere placement in a regular setting without any support is not enough. Interactions between children with and without Down syndrome need to be modelled and fostered. Furthermore, although regularly placed children are generally fairly well accepted by their peers, they are less often seen as 'best' friend. Apparently, in special education, there exist more opportunities for being 'best' friends.

Conclusion Children with Down syndrome learn more academic and language skills in regular education, not only because of selective placement. They are well accepted by their peers. However, peer interactions need to be modelled and fostered. Furthermore, opportunities for the development of intimate 'best' friendships have to be organized explicitly.

Chapter 3 – Multipotential stromal cells, also known as mesenchymal stem cells (MSCs), reside within the bone marrow (BM) and other skeletal and connective tissues. They serve as a reserve of progenitors for continuous cell replacement during normal tissue homeostasis. Furthermore, MSCs are responsible for production of replacement cells lost as a result of acute or chronic injury to bone, cartilage, tendon, meniscus and other connective tissues and are involved in physiological bone remodeling. Loss of MSC function is apparent in the older individual, potentially contributing to the development of osteoporosis and osteoarthritis;

dyskeratosis and Werner's syndromes are also linked to premature aging of skeletal and connective tissues. Telomere attrition has been proposed as a possible mechanism for the loss of MSC function, and may represent a therapeutic target in these diseases.

MSCs can be culture-expanded with relative ease and are already used therapeutically to repair cartilage and bone defects following injury. Also, because of their well-described immunoregulatory capacity, MSC therapy is used to treat patients with graft-versus-host disease. It is therefore very important to monitor telomere dynamics during MSC manufacture, for the purpose of controlling their therapeutic potency and potential for spontaneous transformation.

The authors will outline the current understanding of MSC ageing *in vitro* and observed telomere loss during culture expansion, primarily focusing on MSCs derived from adult tissues (BM, adipose and dental pulp), but also discussing perinatal tissues such as umbilical cord. Various telomere length measurement methods will be described and compared, in terms of their utility and practicality. Safety concerns relating to genetic manipulation of MSCs to extend their proliferation span, and the subsequent quantity of manufactured MSCs will be discussed.

Finally, the authors evaluate *in vivo* MSC ageing compared to that of hematopoietic stem cells highlighting both similarities and differences, which in the authors' view stems from the current lack of robust methods for native MSC isolation directly from skeletal tissues. This combined knowledge should undoubtedly advance the authors' understanding of MSC function during ageing and is likely to contribute to the development of novel therapies for age-related skeletal diseases.

Chapter 4 – The papillomaviruses form a highly diverse group that infects mammals, birds and reptiles. There is a huge veterinary and public health importance associated with these viruses. Because of this, human and bovine papillomaviruses are the most studied viruses of this group. Infection with human papillomavirus (HPV) represents the most important risk factor in the genesis of cervical cancer. Bovine papillomavirus (BPV) infects basal epithelial cells leading to the formation of benign and malignant tumors, when in synergism with action of environmental co-factors. Brazil is a country with continental dimensions; therefore the understanding of the genetic variability and the distribution of these viruses is essential for an active surveillance program and the design of more efficient vaccines and diagnostic methods. In what concerns the association of HPV and cervical cancer, HPV-16 is the most frequent type worldwide, with the exception of Indonesia where the HPV-18 happens to be the most common type. Although the HPV-16 is the most frequent, the second most important type differs between different regions in the world. So, this difference in the incidence of HPV genotypes needs to be elucidated, along with the genetic variability of these isolates that could lead to distinct oncogenic potential among different variants and subtypes. Oncogenic variants of HPV-31 were found and some of the mutations were observed in regions involved in host immune recognition. Changes in amino acid sequences between variants from different geographical regions evidence the possibility of functional divergence between them, suggesting that the variants detected may have different adaptive fitness. On the other hand, twelve BPV types are known so far, but only three of them are associated with cancer. There is a great diversity of BPV types among cattle and high co-infection was found. The presence of infection with multiple viral types could lead to suppression of the immunological system making the clinical state of the animals worse. So, further molecular epidemiological investigations on the incidence and diversity of papillomavirus infection will assist in establishing a more accurate view of the distribution of these viruses. This could help the development of accurate and specific diagnostics, prophylactics and therapeutic requirements.

Chapter 5 – Human telomeres are nucleoprotein complexes located at the end of chromosomes, composed of tandem repeats of the non-codificant DNA sequence TTAGGG. A core of six telomere binding proteins, the shelterin complex, serves to protect telomeric ends and to prevent recognition of chromosome ends as damaged DNA. Both critically short telomeres and direct disruption of the shelterin structure can initiate telomere dysfunction. The enzyme telomerase is a ribonucleoprotein that compensates the telomere reduction by adding new repeats to chromosome ends. Different studies have demonstrated that reduced telomere length (TL) contribute directly to genomic instability, usually found in several types of cancer. Short telomeres and high telomerase activity have been associated with tumorigenesis in patients with hematological neoplasm. Lymphoid malignancies constitute a heterogeneous group of lymphoproliferative disorders that represent distinct disease entities, according to their clinical, morphological, immunophenotypic and genetic features. In this study, the authors analyzed the TL in 262 patients with different B-cell malignancies and its association with genetic characteristics and telomere-associated proteins expression. TL was highly heterogeneous and significantly reduced in B-cell disorders compared to normal controls. Patients with diffuse large B-cell lymphoma secondary to follicular lymphoma (DLBCL-S) had the shortest telomeres, followed by follicular lymphoma (FL). Mantle cell lymphoma (MCL) and *de novo* DLBCL showed similar mean TL, whereas monoclonal gammopathy of undetermined significance (MGUS) and chronic lymphocytic leukemia (CLL) cases displayed the longest TL. Mean TL for multiple myeloma (MM) were intermediate among those entities. Significant shorter TL in MM and CLL patients with chromosomal abnormalities with respect to cases with normal karyotypes were observed ($p<0.03$). In FL, a significant difference between *BCL2* positive and negative cases was found ($p<0.05$). The analysis of telomere-associated proteins expression was performed in MGUS and MM. A significant increase of *TRF1* levels and a lower expression of *TRF2* and *TANK1* in MGUS with respect to MM patients were found ($p=0.006$). The analysis of TL according to *hTERT* expression showed that patients with the highest mRNA values had the shortest TL. These findings are of interest considering the new therapeutic approaches with diverse telomere-targeting inhibitors and suggest that a number of patients with B-cell malignancies may be particularly sensitive to telomere-damage drugs.

Chapter 6 – Telomeres, the ribonucleoprotein complexes located at the chromosomal termini, are committed to cap and protect the endmost chromosomal sequences from being recognized as double strand breaks by the DNA damage repair machinery. Telomeres are synthesized by the functions of the telomerase, a ribonucleoproteic enzyme-complex with reverse transcriptase properties that maintains telomere length homeostasis in eukaryotic cells. In *Ustilago maydis*, a basidiomycetous phytopathogen that infects many plant species of the *Zea* genus, the telomere is composed af a TTAGGG repeated motif, which is commonly found in vertebrates and other fungal species. Both the telomere organization and the characterization of the telomerase catalytic subunit have been characterized in this fungus. Adjacent to the telomere are mosaics of sequences, which are largely species-specific and commonly termed telomere-associated sequences (TAS), The TAS are not fully understood and require further characteriation. *UTASa*, a class of TAS in *U. maydis,* harbors an open reading frame with homology to RecQ enzymes. Those sequences resemble a long non-coding RNA (lncRNA) and exhibit a variable expression pattern during the cell cycle in synchronous cell cultures. The authors propose that this type of RecQ-like sequence may be part of a surveillance mechanism that operates in the absence of telomerase.

Chapter 7 – MicroRNAs (miRNAs) are small (~24 nucleotides) noncoding RNA molecules thought to play an important role in regulating gene expression. Although knowledge of the biological functions of most miRNAs is still limited, miRNAs are thought to regulate the gene expression in various diseases. In the article, the roles of miRNAs in human cancers are examined from the viewpoint of up- and down-regulation. Oncogenic miRNAs are involved in the overexpression/upregulation of cancers, whereas suppressive miRNAs are involved in the underexpression/downregulation of cancers. The oncogenic and suppressive miRNA sequences were first collected from miRBase based on the relations between miRNAs and various cancers in the literature. Statistical analysis of the positional nucleotide occurrence features of miRNAs revealed differences between the positional nucleotide occurrences of oncogenic and suppressive miRNAs. A miRNA gene-silencing score was then defined on the basis of the higher and lower levels of the statistical significances of positional nucleotides. Since the miRNA scores were closely related to miRNA frequencies, a method using the scores and nucleotide frequencies to distinguish whether a new miRNA is oncogenic or suppressive is proposed.

Chapter 8 – Analyzing and generating music material has been a field of intense research, combining scientific disciplines from computational intelligence and signal processing or computational music analysis. This chapter reviews the application of methods that utilize genetic programming (GP) along with other methods for tackling several tasks that pertain to audio and music analysis and synthesis. The focus of this overview is in the evolution of the methodologies themselves and therefore, the discussed papers are mostly presented in a chronological order. In musical audio analysis, GP–based methodologies were developed to performed musical instrument identification, i.e., to recognize the instruments that are included in musical recordings. For audio synthesis, GP has been primarily utilized to evolve combinations of sound generating blocks or simple mathematical functions that shape waveforms. Symbolic music composition has been mostly approach by evolving simple programs that modify short melodic segments and concatenate them to compose a complete melodic part. For the sound and music synthesis methodologies, the main concern revolves around formalizing efficient evaluation techniques for the computation of the fitness value. On the one hand, interactive evaluation, where the evolution is driven through user ratings, incorporates the phenomenon of user fatigue, according to which the users provide small numbers of potentially contradictory evaluations. On the other hand, efficient automatic evaluation requires the development of qualitative music and sound descriptors, a fact that constitutes by itself a deep research subject. The chapter concludes by addressing issues concerning the utilization of GP towards approaching effective solutions to define proper music and sound features, as well as to alleviate user fatigue.

In: Advances in Genetics Research. Volume 12
Editor: Kevin V. Urbano
ISBN: 978-1-63321-351-7

Chapter 1

THE "S" ALLELE OF THE SEROTONIN TRANSPORTER IS NOT ASSOCIATED WITH MAJOR DEPRESSION IN A SAMPLE OF VETERANS

Jack R. Cornelius,*[*] *Gretchen L. Haas, Gerald Goldstein, Barbara Hanusa, Jon D. Walker, Lauren J. Fox and James Ferrell
VISN 4 Mental Illness Research, Education and Clinical Center (MIRECC),
VA Pittsburgh Healthcare System, Pittsburgh, PA, US

Presented in part at the 23rd Annual Meeting and Symposium of the American Academy of Addiction Psychiatry (AAAP), Aventura, Florida, December 6-9, 2012.

ABSTRACT

The results of some studies suggest that the serotonin transporter-linked polymorphic region (5-HTTLPR) short (S) allele, relative to the long (L) allele, is associated with risk for Major Depressive Disorder (MDD), and thus serves as a biomarker for MDD, while results from other studies do not support that conclusion. Persons with an S allele demonstrate a 2- to 2.5 fold decrease in serotonin transcription rate compared to the L-allele, which may increase their risk for MDD. Differences in study populations may help explain the differences in findings between those meta-analyses. To date, there have been no published reports which have addressed the possible association between the S allele and MDD among military veterans. This manuscript describes a first study to assess the possible association of the S allele with MDD among a study population of veterans in treatment for a substance use disorder. We hypothesized that the S allele would be associated with MDD in our study sample. Subjects signing informed consent were 101 Veterans recruited from VA behavioral health and substance use treatment clinics in the VA Pittsburgh Healthcare System, and 91 of those subjects were genotyped for 5-HTTLPR polymorphisms. The study sample from whom genetic material was collected included 82 males and 9 females, of whom 53 were white, 38 were black, and one was

[*] Telephone 412-246-5186. E-mail: corneliusjr@upmc.edu.

"other". Fifty-four members of the study sample (59%) met DSM-IV criteria for an MDD on the SCID. Forty-five of the subjects demonstrated one or two S alleles, while 46 did not do so. The presence of the S allele of the serotonin transporter was not found to be significantly associated with the diagnosis of major depressive disorder in our sample (Chi-square=0.1.63, df=1, p=0.199). That finding, in combination with other recent negative findings from other researchers involving non-veterans, raises questions regarding the clinical utility of utilizing genetics tests involving the assessment of the alleles of the serotonin transporter as a possible biomarker for MDD.

1. INTRODUCTION

Serotonin (5-HT) is a monoamine neurotransmitter with multiple sites of action, which can affect mood, sensory processing, cognition, and sleep (Covault et al., 2007). Serotonergic neurotransmission has been implicated in the pathogenesis of mood disorders such as major depressive disorder and a variety of other disorders (Nemeroff & Owens, 2002; Nelissery et al., 2003. The 5-HT transporter protein (also called the 5-HTT) plays a key role in regulating the serotonergic system by regulating the reuptake of serotonin (5-HT) from the synaptic cleft following synaptic release (Hranilovic et al., 2004; Otte et al., 2007; Kilpatrick et al., 2007). The 5-HTT protein is encoded by a single gene, the serotonin transporter gene (SLC6A4), a gene in the regulatory pathway for serotonin. The serotonin transporter gene has a 44-bp insertion/deletion polymorphism in its regulatory region, known as the 5-HTT linked promoter region (5-HTTLPR). This deletion/insertion polymorphism results in two common alleles, the long allele (L) and the short allele (S) (Tartter and Ray, 2011). The S allele is associated with a 2- to 2.5 fold decrease in 5-HTT transcription rate compared to the L allele, resulting in less efficient serotonin reuptake and decreased expression of the serotonin transporter (Lesch et al., 1996; Heils et al., 1996). The presence of the S allele has been reported to be associated with an increased susceptibility to major depression (Fein et al., 2005; Zalsman et al., 2006).

The results of some studies suggest that the serotonin transporter linked polymorphic region (5-HTTLPR) short (S) allele, relative to the long (L) allele, is associated with risk for Major Depressive Disorder (MDD), and thus may serve as a biomarker for MDD (Carver et al., 2008; Wray et al., 2009. Similarly, results from Nellissery et al. (2003) suggest that the S allele is associated with Major Depression among subjects with co-occurring alcohol dependence.

However, results from other studies do not support the association between the S allele and major depressive disorder (Minov et al., 2001; Anguelova et al., 2003; Risch et al., 2009). Those mixed findings concerning the effect of the S allele on major depressive disorder may be due to small genetic effects, or may be due to differences in study samples. No previous studies of this allele have been conducted in a study sample of veterans. To help resolve the mixed findings, and to further address the possible association of the S allele with major depression among individuals with co-occurring alcohol and/or drug dependence, the current study was conducted in a sample of veterans. To our knowledge, the present study is a first published study to assess the effect of the S allele of the serotonin transporter gene in the etiology of major depressive disorder in a study sample involving veterans.

2. Method

2.1. Subjects

Subjects were Veterans recruited from the outpatient behavioral health services of the VA Pittsburgh Healthcare System (VAPHS). Before entry into this protocol, the study was explained to all participants, and written informed consent was obtained from all subjects after all procedures had been fully explained. The study was approved by the VAPHS Institutional Review Board. This study was conducted at the Highland Drive Campus of the VAPHS. Subjects were recruited for participation in the study through referrals from the outpatient behavioral health clinic and substance abuse treatment programs at the VAPHS and by flyers distributed at the VA and in the community. Study participants were required to be over the age of 18 years of age at recruitment into the study. Subjects were recruited without regard to gender, race, or ethnicity. Potential subjects were excluded from participation for any psychosis-related diagnosis, such as schizophrenia, and for evidence (by research interview) of any medical condition that would confound determination of any Axis I diagnosis. The study was typically completed in one to two sessions. Subjects were compensated $20 for participation in the study. Blood was also collected for genotyping as part of this study.

2.2. Diagnostic Instrument

The SCID

Psychiatric diagnoses (including the diagnosis of PTSD) were determined using the Structured Clinical Interview for the DSM-IV (SCID) (First et al., 1997), administered by experienced clinical interviewers with specialized training in SCID administration. The SCID has been shown to demonstrate good reliability, validity, and clinical utility (First et al., 2008).

2.3. Genotyping

Two purple top EDTA 10 cc vacutainer tubes were drawn for each subject. The blood samples were transported to Dr. Ferrell's Genetics laboratory at the Graduate School of Public Health at the University of Pittsburgh for genotyping. All samples were coded so as not to allow persons unassociated with the study to connect the blood samples to a specific individual. The only person with access to the linkage code between the participants and their respective blood samples was the research coordinator, and the linkage codes were securely stored at all times. DNA was isolated using the PureGene kit and genotyped as previously described (Flory, Manuck, Ferrell et al., 1999**.** In brief, processing of the sample occurred by centrifugation, which separated the samples into three distinct materials: plasma, a buffer portion that contained the DSN, and red cells. Allelic variation was determined by PCR DSA amplification, and alleles were designated S (484bp) and L (528bp) by direct comparison to control samples run on the same gel. The plasma and the red cells were extracted and destroyed immediately. Samples were refrigerated in a secure temperature-controlled environment until enough samples had been gathered to perform a genotyping analysis. Following the completion of the genotyping, the genetic materials and associated electronic data concerning the genotyping were transferred back to the VA in a secure manner for storage.

2.4. Statistical Analysis

Descriptive statistics were calculated for all variables. Between-group comparisons of categorical variables were performed using chi-square tests, corrected for continuity. All tests were two-tailed. An alpha level of less than or equal to 0.05 was used in the study. All analyses were conducted using the Statistical Package for the Social Sciences, Version 15.0 (Norusis, 1992). Logistic analyses were planned as a follow-up to the chi-square analyses, to be conducted if any of the findings from the chi-square tests were statistically significant. However, none of the results from the chi-square tests were statistically significant, so logistic analyses were not performed.

3. Results

This study recruited 107 potential subjects. DNA from a total of 91 veterans was collected and genotyped for 5-HTTLPR polymorphisms, in order to compare the distribution of S and L alleles. These subjects included 82 males (90%) and 9 females (10%); of whom 52 were white (57%), 38 were black (42%), and one (1%) % self-identified as "other". Fifty-three members of the study sample (58%) met DSM diagnostic criteria for MDD. Sixty-three members of the study sample (69%) met DSM diagnostic criteria for an alcohol use disorder (alcohol dependence or alcohol abuse), and 59 members of the study sample (65%) met DSM diagnostic criteria for a substance use disorder other than an alcohol use disorder. The presence of the S allele was not found to be significantly associated with any demographic variable.

The presence of the S allele was not found to be significantly associated with the diagnosis of MDD in our sample (Pearson chi-square=0.174, df=90, p=0.677). In addition, the presence of the S allele was not found to be significantly associated with the diagnosis of AUD in our sample (Pearson chi-square=0.17, df=90, p=0.68) or with the diagnosis of SUD in our sample (Pearson chi-square=0.00, df=90, p=1.00).

Conclusion

The findings of this study did not confirm our hypothesis that the S allele of the serotonin transporter gene would be associated with the diagnosis of major depressive disorder. Instead, the findings of this study demonstrate that the presence of the lower expressing allele (the S allele) of the serotonin transporter gene (5-HTTLPR) is not associated with the presence of major depressive disorder among our sample of veterans. There are mixed findings in the literature regarding the association of 5-HTTLPR genotypes and major depressive disorder. Some studies have found that the S allele is associated with major depressive disorder, as reported by Zalsman et al. (2006). Similarly, Nellissery et al. (2003) found that the S allele is associated with major depression among persons with alcohol dependence. Hoefgen et al. (2005) found a small but statistically significant effect of 5-HTTLPR serotonin transporter polymorphism and major depressive disorder, but also concluded that demonstrating that association required a very large and homogenous study sample, such as their own study sample involving only subjects with a German descent. However, others have not found an association between 5-HTTLPR polymorphisms and the presence of major depression (Minov et al., 2001). Studies from a number of non-American study populations_have also shown no association between 5-HTTLPR polymorphisms and mood disorders, such as studies involving Japanese subjects (Kunugi et al., 1997; Ohara et al., 1998), Spanish subjects (Gutierrez et al., 1998), French or German Caucasians (Hoehe et al., 1998), Ashkenazi or non-Ashkenazi Jews (Frisch et al., 1999), or Brazilians (Oliveira et al., 2000). Also, a meta-analysis found no association between 5-HTTLPR genotype and major depression (Anguelova et al., 2003). Similarly, another larger more recent meta-analysis involving 14,250 participants found no association between 5-HTTLPR genotype and major depressive disorder or stressful life events or the interaction between those factors (Risch et al., 2009). Thus, the results of our current study are consistent with the findings of the majority of previous studies, particularly the more recent study.

The reason for these discrepant results from different studies concerning the association between 5-HTTLPR genotype and major depression is unclear, but could involve small effect size or different ethnic distributions of study subjects (Minov et al., 2001). However, the results of those studies, in combination with the results of our current study, suggest that the 5-HTTLPR polymorphism is not a major susceptibility factor in the etiology of major depression (Minov et al., 2001; Risch et al., 2009). That finding is consistent with the result of a large recent British meta-analysis that evaluated the effect of a variety of genetic polymorphisms in the etiology of major depression (Lewis et al., 2010). That British study concluded that individual gene contributions to depression result in at most only minor effects, despite the fact that studies of major depression in twins and families have shown moderate to high heritability for major depressive disorder, so large pooled analyses would be required to identify any possible individual gene contributions. Thus, it appears that the genetic contribution to major depression is attributable to multiple genes each of small effect, as had been noted by Minov et al. (2001).

The findings of this study also demonstrate that the presence of the S allele of the serotonin transporter gene is not associated with the presence of alcohol use disorder or with the presence of substance use disorder among our sample of veterans. There are mixed findings in the literature regarding the association of 5-HTTLPR genotypes and alcohol use disorders. One study found that the S allele is associated with alcohol dependence (Feinn, Nellissery, & Kranzler, 2005). A few other studies have provided evidence of an allelic association with alcohol dependence (Hammoumi et al., 1999, Lichtermann et al., 2000). However, the results of several other studies failed to show a significant association of 5-HTTLPR alleles with alcohol dependence (Edenberg et al.., 1998; Gelenter et al., 1997; Gorwood et al., 2000; Ishiguro et al., 1999, Matsushita et al.,. 2001; Preuss et al., 2001). A meta-analysis across 22 studies found only a small (though significant) association between the S allele and alcohol dependence (McHugh et al., 2010). Furthermore, that meta-analysis noted a greater number of studies showing null results published in more recent years. Indeed, that meta-analysis did not demonstrate a significant association between 5-HTTLPR and alcohol dependence across studies published since 2005. Consequently, that meta-analysis concluded that the effect size of that association was moderated by study sample size and publication year, and may be the result of publication bias. The authors of that meta-analysis recommended that the results suggesting an association between 5-HTTLPR and alcohol dependence be interpreted with caution. The results of previous studies, in combination with the results of our current study, suggest that the 5-HTTLPR polymorphism is not a major susceptibility factor in the etiology of alcohol use disorders.

In summary, the results of the current study replicate and extend findings from non-veteran populations. The authors of the current study agree with the conclusion of Nellissery et al. (2003), that despite substantial evidence of the involvement of serotonin in both major depression and in alcohol use disorders, findings regarding the possible association of 5-HTTLPR alleles with either of those disorders are equivocal, and there is no clear evidence to date has been shown supporting that association among veteran populations.

Limitations of the current study include its modest sample size and possible population stratification. Also, since the sample for the current study was drawn from persons seeking treatment in substance use clinics, the results of the current study may not generalize to all veterans. Further studies are warranted to clarify the association of the 5-HTTLPR polymorphism and diagnoses of major depression and alcohol use disorder among veteran and non-veteran populations. Further studies are also warranted to assess whether the 5-HTTLPR polymorphism serves as a predictor of treatment response for depressive symptoms or for level of alcohol or substance use.

Authors' Note

The contents of this manuscript do not represent the views of the Department of Veterans Affairs or the United States Government.

The authors acknowledge the assistance of Maribeth Wesesky in the preparation of this manuscript.

Acknowledgments

This material is based upon work supported by the Department of Veterans Affairs, Veterans Health Administration, the VAPHS Office of Research and Development and the VISN 4 Mental Illness Research, Education, and Clinical Center (MIRECC, Director: D. Oslin; Pittsburgh Site Director: G. Haas), VA Pittsburgh Healthcare System. This research was also supported in part by grants from the National institute on Drug Abuse (P50 DA05605, R01 DA019142, R01 DA14635, K02 DA017822, R01 DA034072, and the NIDA Clinical Trials Network); and from the National Institute on Alcohol Abuse and Alcoholism (R01 AA013370, R01 AA015173, R01 AA014357, R01 AA13397, R21 AA022123, R21 AA022863, K24 AA15320, and K02 AA000291.

REFERENCES

Anguelova, M., Benkelfat, C., Turecki, G. (2003). A systematic review of association studies investigating genes coding for serotonin receptors and the serotonin transporter, II: suicidal behavior. *Molecular Psychiatry, 8:*646-653.

Carver, C.S., Johnson, S.L., & Joormann, J. (2008). Serotonergic function, two-mode models of self-regulation, and vulnerability to depression: What depression has in common with impulsive aggression. *Psychological Bulletin 134*:912-943.

Edenberg, H.J., Reynolds, J., Koller, D.L., Beglieter, H., et. al. (1998). A family-based analysis of whether the functional promoter alleles of the serotonin transporter gene HTT affect the risk for alcohol dependence. *Alcohol, Clinical and Experimental Research, 22:*1080-1085.

Feinn, R., Nellissery, M., Kranzler, H.R. (2005). Meta-analysis of the association of a functional serotonin transporter promoter polymorphism with alcohol dependence. *American Journal of Medical Genetics Part B, 133B:*79-84.

Frisch, A., Postilnick, D., Rockah, R., Michaelovsky, E., Postilnick, S., Birman, E., Laor, N., el al. (1999). Association of unipolar major depressive disorder with genes of the serotonergic and dopaminergic pathways. *Molecular Psychiatry, 4:*389-392.

Gelernter, J., Kranzler, H., Cubells, J. (1997). Serotonin transporter protein (SLC6A4) allele and haplotype frequencies and linkage disequilbria in African- and European-American and Japanese populations and in alcohol-dependent subjects. *Human Genetics, 101:*243-246.

Gorwood, P., Batel, P., Ades, J., Hamon, M., Boni, C. (2000). Serotonin transporter gene polymorphisms, alcoholism, and suicidal behavior. *Biological Psychiatry, 48:*259-264.

Gutierrez, B., Pintor, L., Gasto, C., Rosa, A., Bertranpetit, J., Vieta, E., Fananas, L. (1998). Variability in the serotonin transporter gene and increased risk for major depression with melancholia. *Human Genetics, 103:*319-322.

Hammoumi, S., Payen, A., Favre, J.D., Balmes, J.L., Benard, J.Y., Husson, M. Ferrand, J.P., Martin, J.P., Daoust, M. (1999). Does the short variant of the serotonin transporter linked polymorphic region constitute a marker of alcohol dependence? *Alcohol, 17:*107-112.

Heils, A., Teufel, A., Petri, S., Stober, G., Riederer, P., Bengel, D., Lesch, K.P. (1996). Allelic variation of human serotonin transporter gene expression. *Journal of Neurochemistry, 66*: 2621-2624.

Hoefgen, B., Schulze, T.G., Ohlraun, S. et al. (2005). The power of sample size and homogenous sampling: association between the 5-HTTLPR serotonin transporter polymorphism and major depressive disorder. Biological Psychiatry, *57*:247-251.

Hoehe, M.R., Wendel, B., Grunewald, I., Chiaroni, Pl, Levy, N., Morris-Rosendahl, D. et al. (1998). Serotonin transporter (5-HTT) gene polymorphisms are not associated with susceptibility to mood disorders. *American Journal of Medical Genetics*, *8*1:1-3.

Hranilovic, D., Stefulj, J., Schwab, S., Borrmann-Hassenback, M, Albus, M., Jernej, B., Wildenauer, D. (2004). Serotonin transporter promoter and intron 2 polymorphisms: relationship between allelic variants and gene expression. *Biological Psychiatry*, 55:1090-1094.

Ishiguro, H., Saito, T., Akazawa, S., Mitushio, H., Tada, K., Enomoto, M., Mifune, H., Toru, M., Shibuya, H., Arinami, T. (1999). Association between drinking related antisocial behavior and a polymorphism in the serotonin transporter gene in a Japanese population. *Alcohol Clinical and Experimental Research, 23:*1281-1284.

Kilpatrick, D.G., Koenen, K.C., Ruggiero, K.J., Acierno, R., Galea, S., Resnick, H.S., Roitzsch, J., Boyle, J., Gelernter, J. (2007). The serotonin transporter genotype and social support and moderation of posttraumatic stress disorder and depression in hurricane-exposed adults. *American Journal of Psychiatry, 164*:1693-1699.

Kunugi, H., Hattori, M., Kato, T., Tatsumi, M.Sakai, T., Sasaki, T., Hirose, T., Nanko, S. (1997). Serotonin transporter gene polymorphisms: ethnic difference and possible association with bipolar affective disorder. *Molecular Psychiatry, 2:*457-462.

Lesch, K.P., Bengel, D., Heils, A., Cabol, S.Z., Greenberg, B.D., Petri, S., Benjamin, J., Mueller, C.R., Hamer, D.H. Murphy, D.L. (1996). Association of anxiety-related traits with a polymorphism in the serotonin transporter gene regulatory region. *Science, 274*:1527-1531.

Lewis, C.M., Ng, M.Y., Butler, A.W., Cohen-Woods, S., Uher, R. et al. (2010). Genome-wide association study of major recurrent depression in the U.K. population. *American Journal of Psychiatry, 167*:949-957.

Lichtermann, D., Hranilovic, D., Trixler, M., Franke, P., Jernej, B., Delmo, C.D. et al. (2000). Support for allelic association of a polymorphic site in the promoter region of the serotonin transporter gene with risk for alcohol dependence. *American Journal of Psychiatry, 157*:2045-2047.

Lotrich, F.E., Pollock, B.G., Ferrell, R.E. (2003). Serotonin transporter promoter polymorphism in African Americans. Allele frequencies and implications for treatment. *American Journal of Pharmacogenetics, 3:*145-147.

Matushita, S., Yoshino, A., Murayama, M., Kimura, M., Muramatsu, T., Higuchi, S. (2001). Association study of serotonin transporter gene regulatory region polymorphism and alcoholism. *American Journal of Medical Genetics, 105:*446-450.

McHugh, R.K, Hofmann, S.G., Asnaani, A., Sawyer, A.T., Ottoa, M.W. (2010). The serotonin transporter gene and risk for alcohol dependence: A meta-analytic review. *Drug and Alcohol Dependence, 108:*1-6.

Minov, C., Baghai, T.C., Schule, C., Zwanzher, P., Schwarz, M.J., Zill, P., Rupprect, R., Bondy, B. (2001). Serotonin-2A-receptor and –transporter polymorphisms: lack of association in patients with major depression. *Neuroscience Letters, 303:*119-122.

Nellissery, M., Feinn, R., Covault, J., Gelernter, J., Anton, R., Pettinati, H., Moak, D., Mueller, T., Kranzler, H. (2003). Alleles of a functional serotonin transporter promoter polymorphism are associated with major depression and alcohol dependence. *Alcohol Clinical and Experimental Research, 27:*1402-1408.

Nemeroff, C.B., Owens, M.J. (2002). Treatment of mood disorders. *Nature Neuroscience, 5(November supplement)*:1068-1070.

Norussis, M.J. (1992). Statistical Package for the Social Sciences. McGraw-Hill, New York, NY.

O'Hara, K., Nagal, M., Tsukamoto, T., Tani, K., Suzuki, Y., O'Hara K. (1998). Functional polymorphism in the serotonin transporter promoter at the SLC6A4 locus and mood disorders. *Biological Psychiatry, 44:*550-554.

Oliveira, J.R., Carvalho, D.R., Pontual, D., Gallindo, R.M., Sougey, E.B., Genil, V. et al. (2000). Analysis of the serotonin transporter polymorphism (5-HTTLPR) in Brazilian patients affected by dysthymia, major depression and bipolar disorder. *Molecular Psychiatry, 4*:348-349.

Otte, C., McCaffery, J., Ali, S., Whooley, M.A. (2007). Association of a serotonin transporter polymorphism (5-HTTLPR) with depression, perceived stress, and norepnephrin in patients with coronary disease: The heart and soul study. *American Journal of Psychiatry, 164*: 1379-1384.

Preuss, U.W., Koller, G., Soyka, M., Bondy, B. (2001). Association between suicide attempts and 5-HTTLPR-S-allele in alcohol-dependent and control subjects. Further evidence from a German alcohol-dependent inpatient sample. *Biological Psychiatry, 50:*636-639.

Risch, M., Herrell, R., Lehner, T., Liang, K.Y., Eaves, L., Hoh, J., Griem, A., Kovacs, M., Ott, J., Merikangas, K.R. (2009). Interaction between the serotonin transporter gene (T-HTTLPR), stressful life events, and risk of depression. *JAMA, 301*:2462-2471.

Wray, N.R., James, M.R., Gordon, S.D., Dumenil, T., Ryan, L, Coventry, W.L., Statham, D.J., Pergadia,M.L., Madden, P.A., Heath, A.C., Montgomery, G.W., Martin, N.G. (2009). Accurate, large scale genotyping of 5HTTLPR and flanking single nucleotide polymorphisms in an association study of depression, anxiety, and personality measures. *Biological Psychiatry, 66*:468-476.

Zalsman, G., Huang, Y.Y., Oquendo, M.A. et al. (2006). Association of a triallelic serotonin transporter gene promoter region (5-HTTLPR) polymorphism with stressful life events and severity of depression. *American Journal of Psychiatry, 163*:1588-1593.

In: Advances in Genetics Research. Volume 12
Editor: Kevin V. Urbano

ISBN: 978-1-63321-351-7

Chapter 2

EFFECTS OF REGULAR VERSUS SPECIAL SCHOOL PLACEMENT ON STUDENTS WITH DOWN SYNDROME: A SYSTEMATIC REVIEW OF STUDIES

***Gert de Graaf*[1,2,*], *Geert van Hove*[1] and *Meindert Haveman*[3]**
[1]Department of Orthopedagogics, Ghent University, Belgium
[2]Dutch Down Syndrome Foundation, Meppel, the Netherlands
[3]Department of Rehabilitation Sciences, University of Dortmund, Germany

ABSTRACT

Background Since the 1980s, more and more children with Down syndrome are in regular education. Parents and schools expect social and cognitive advantages.

Method This systematic review of studies on the effects of school placement of students with Down syndrome, with special reference to self-help skills, language, academics and social functioning, is based on the following criteria. Firstly, studies were published in the period 1970-2010. Secondly, any study with a direct comparison between placements in developmental or social outcomes was included. However, studies with a very small sample size (n<3) were excluded. Thirdly, non-comparative studies were included if in the study the acceptance of regularly placed children with Down syndrome by their own classmates was evaluated. Single case studies were excluded. Finally, studies were published in English, Dutch, German, French, Italian, Portuguese, Spanish, Norwegian, Swedish, Danish or Greek. To detect studies, comprehensive sources such as Picarta, Google, Medline, ERIC and Science Direct were used. In addition, a request for relevant research reports was sent out to all member organisations of the European Down Syndrome Organisation (EDSA).

Results The literature search yielded 133 potentially relevant studies, of which 53 met the inclusion criteria. In 26 studies, regularly and specially placed children with Down syndrome were compared and in another 27 studies, the acceptance of regularly placed children with Down syndrome by their classmates was evaluated. In nine of the 26 comparative studies, no attempt was made to disentangle the effect of selective placement from the effect of differences in stimulation between settings. However, in fifteen studies

---

* Corresponding author: Email: Graaf.Bosch@ziggo.nl or gertdegraaf@downsyndroom.nl

researchers did statistically correct for the effect of important other child and/or family variables that could have an impact on development. Furthermore, four studies can be considered to be natural experiments in which school placement was not determined by child characteristics but by geographical area and/or generation. Results show that regular classroom placement yields a better development of language and academic skills, even after the effect of selective placement has been taken into account. As regards self-help skills, under the same condition, there seem to be no differences between both types of school. Social functioning shows a mixed image. For social network, behaviour, and self-competence, no differences at all or small positive differences for regularly placed children were found. However, most studies also highlight that mere placement in a regular setting without any support is not enough. Interactions between children with and without Down syndrome need to be modelled and fostered. Furthermore, although regularly placed children are generally fairly well accepted by their peers, they are less often seen as 'best' friend. Apparently, in special education, there exist more opportunities for being 'best' friends.

Conclusion Children with Down syndrome learn more academic and language skills in regular education, not only because of selective placement. They are well accepted by their peers. However, peer interactions need to be modelled and fostered. Furthermore, opportunities for the development of intimate 'best' friendships have to be organized explicitly.

Keywords: Down syndrome, education, disability, inclusion, inclusive education

INTRODUCTION

In many developed countries the right to public school education for children with Down syndrome was not established until the 1970s. With the exception of Italy, a country adopting a full inclusive educational system as early as 1977, in practical terms this newly acquired right to education almost inevitably meant special school placement. In the 1970s, and in most countries even in the early 1980s, very few students with Down syndrome entered regular education.

However, since the mid 1980s in many countries, including the UK (Cuckle, 1997), Australia (Bochner and Pieterse, 1996) and the Netherlands (de Graaf, 2007a; Scheepstra, 1998), more and more children with Down syndrome are entering regular schools. For children with Down syndrome, the parent's choice for more inclusion has been and still is the driving force for changes in educational placements. Dutch studies show that parents with children with Down syndrome choose regular schools for social, educational and ethical reasons (de Graaf, 1998; Pijl and Scheepstra, 1998; Poulisse, 2002).

The consequences of the social principle of inclusive education, the expectation that in the long term regular school placement of children with disabilities will lead to a more open and tolerant society, are rather difficult to research in any direct way. However, one could argue that a more tolerant society derives from positive experiences in childhood, such as children with disabilities being accepted by peers. The extent to which this is really happening in the schools can be researched. Consequently, an important research topic is the question whether regular school placement really leads to the social and cognitive advantages these parents (and those regular schools which agree to take the child) are expecting and hoping for. Does regular placement of students with Down syndrome really lead to better self-help skills, language development, academics and social functioning?

Earlier Reviews on the Effect of Placement for Children with Disabilities

Studies into the effects of special versus regular placement for children with disabilities date back into the 1930s (Carlberg and Kavale, 1980). Three meta-analyses (Baker, Wang, and Walberg, 1994/95; Carlberg and Kavale, 1980; Wang and Baker, 1985/86) conclude that placement in a regular class leads to small to moderate positive effects on cognitive and social development of students with intellectual disabilities (ID). However, in a recent review, Lindsay (2007) states that there is not enough methodologically sound research to prove the superiority of inclusive education for students with disabilities. However, Lindsay's review is rather limited in scope, only reviewing articles in eight relevant journals published in the period 2001-2005. In contrast, Ruijs and Peetsma (2009) reviewed a larger body of research and conclude that inclusive education leads to neutral to positive effects on cognitive and socio-emotional development. The academic achievement of students with special educational needs in inclusive classes is comparable or even better to that of their counterparts in non-inclusive classes. However, most studies in these reviews focus on students with mild disabilities, including many studies on students with specific learning difficulties. Only one review by Freeman and Alkin (2000) specifically aims at research on students with ID, ranging from mild to severe ID. These reviewers state that many of these studies suffer from methodological shortcomings. Nevertheless, they conclude that the research seems to support the notion that students with ID develop more academic and social competence in regular schools than in special education, also if comparing students of similar intelligence. However, we cannot assume that the results of these reviews will automatically apply to students with Down syndrome as well. We will have to focus on Down syndrome specific research.

Earlier Down Syndrome Specific Reviews

In contrast to our review of international research, earlier reviews on the effect of regular versus school placement on students with Down syndrome were narrower in scope, and in particular covered only studies published in English. Cunningham, Glenn, Lorenz, Cuckle, and Shepperdson (1998) identified only four outcome studies at that time, three in the UK (Casey, Jones, Kugler, and Watkins, 1988; Philps, 1992; Sloper, Cunningham, Turner, and Knussen, 1990) and one in the USA (Fewell and Oelwein, 1990). These four studies are included in our review. Buckley and Bird (2000) identified another five comparative studies from the UK (Beadman, 1997; Dew-Hughes and Blandford, 1998; Gould, 1998; Laws, Byrne, and Buckley, 2000; and the study later published as Buckley, Bird, Sacks, and Archer, 2006). In addition they reviewed two non-comparative studies on social acceptance and interactions of students with Down syndrome in mainstream schools (Laws, Taylor, Bennie, and Buckley, 1996; Quail, 2000). All of these studies are included in our review as well, with the exception of Dew-Hughes and Blandford (1998), as in presenting their results no distinction was made between students with and without Down syndrome and, moreover, out of 12 participants only two had Down syndrome.

Both Cunningham et al. (1998) and Buckley and Bird (2000) conclude that regular classroom placement appears to support academic skill development in children with Down syndrome.

METHOD

Criteria for Inclusion and Exclusion of Studies

Our review covers the period 1970-2010. Studies were published in English, Dutch, German, French, Italian, Portuguese, Spanish, Norwegian, Swedish, Danish or Greek. We included in our review any study in which a comparison was made between regularly and specially placed children with Down syndrome in their development or functioning. Subsequently, the material could be divided in results regarding self-help skills, language, academics and/or social functioning. However, we excluded studies in which only one regularly placed child was compared with only one specially placed other child, as this approach will yield rather idiosyncratic results. Furthermore, most comparative studies contain a regularly placed group and a specially placed control group. However, sometimes a comparison is made with a reference group outside the specific study. In that case, this study was only included if adequate and precise information on this reference group was presented.

We not only included studies on the effect of *school* placement, but also studies on the effect of regular versus special *preschool* placement, as in different countries the moment children enter the school system can vary. For instance in the Netherlands children enter the school system at age 4, while in many other countries children between 4 and 6 are still in preschool.

To avoid publication bias we included not only studies published in scientific journals, but also studies published as book, conference paper, PhD or master's thesis. We searched for articles published since 1970, because worldwide only since the 1970s, and in many countries even later, children with Down syndrome were granted the right to public school education. As a result, before the 1970s only few entered the educational system at all, and regular school placement must have been extremely rare.

Finally, we included not only studies targeted solely on pupils with Down syndrome, but also studies in which children with Down syndrome formed a substantial subgroup of a mixed research population. However, the latter were only included if the researchers presented their results separately for pupils with and without Down syndrome, or if they had checked that their results were not different for pupils with and without Down syndrome, or if a vast majority of the children in the study had Down syndrome.

In addition, studies without a direct comparison between placements were included if in the study the acceptance of regularly placed children with Down syndrome by their classmates was evaluated. Any study measuring this acceptance (through interviews or questionnaires with children, parents or teachers, sociometric methods or direct observation) was included, with the exception of: 1) single case studies; 2) studies targeted at investigating ways of social support in the regular school, but not actually evaluating the extent of acceptance by peers in any systematic way; 3) studies measuring attitudes of children in mainstream schools towards children with Down syndrome in general, however without them having any actual classroom contact with children with Down syndrome.

Literature Search

To reach a complete overview of studies on this issue over the period 1970-2010, a literature search was conducted in Picarta, Google, Medline, ERIC and Science Direct, using different combinations of relevant entries (Down syndrome, trisomy-21, integration, inclusion, inclusive, education, disability, school; and also equivalents in Dutch, German and French). If in a publication authors referred to other studies on this topic, we tried to obtain these as well.

To get an even broader view, beyond these language boundaries, a request for relevant research reports was sent out on the mailing list of the member organisations of the European Down Syndrome Organisation (EDSA). In this way, it was possible to reach Board members of more than 40 syndrome specific European organisations in more than 25 countries. However, this yielded response only from Germany, France, Poland, Belgium and Malta. The organisations from these first three countries reported that to their knowledge no relevant Down syndrome specific studies on regular school placement had been conducted in their country. The organisation from Belgium referred to one study; however, this article already had been found through the literature search and, moreover, didn't meet the inclusion criteria. From Malta one study was derived, also not meeting the inclusion criteria.

Through the Down syndrome Organisation, we were brought into contact with researchers with experience in educational research on children with Down syndrome from Norway (Anne-Stine Dolva) and Italy (Renzo Vianello). Dolva (email to the first author 9-8-2011) reported that, apart from her own dissertation, to her knowledge no relevant Down syndrome specific educational studies had been conducted in Norway, Sweden and Denmark. Vianello (email to the first author 14-7-2011) sent two articles from Italy, one on attitudes of teachers, parents and peers towards students with Down syndrome and another one on the development of students with Down syndrome. Vianello stated that in the Italian system all children with Down syndrome go to regular schools, making it impossible to conduct studies on the effects of regular placement with a specially placed control group.

Subsequently, in addition to the request to the European Down syndrome Organisation, another literature search was conducted in Google with combinations of the afore mentioned entries, but now in Italian, Portuguese, Spanish, Norwegian, Swedish, Danish and Greek. This yielded educational studies on children with Down syndrome from Latin America, Italy and Greece. However, most did not meet our inclusion criteria.

RESULTS

The literature search yielded 133 potentially relevant educational studies on students with Down syndrome from 23 different countries. 70 of these studies had been conducted in the period 2000-2010, 39 in the 1990s, 22 in the 1980s and only 2 in the 1970s, confirming the fact that in the 1970s, integration in regular education of children with Down syndrome must have been extremely rare.

This total of 133 potentially relevant studies contained 31 comparative studies from 8 different countries and 102 non-comparative studies from 23 different countries. Applying the exclusion criteria, the total number of included studies was reduced to 53 from 12 different countries, with 26 comparative studies from 6 different countries (Argentina, Australia, Ireland, the Netherlands, UK, USA) and 27 non-comparative studies from 11 different countries (Australia, Colombia, Guatemala, Ireland, Italy, the Netherlands, New Zealand, Norway, South Africa, UK, USA). Consequently, bearing in mind that four decades of international research in 11 different languages were reviewed, the picture that emerges is the relative scarcity of studies on the topic.

In table 1, in order of year of publication, all comparative studies included are presented. In table 2 the non-comparative studies included are presented. If the same study was reported in different publications, for instance as a thesis and as articles, the publication with most details is referred to in the first column in the tables 1 and 2.

Table 3 presents studies excluded, explaining the reasons why. Out of the pool of potentially relevant comparative studies, one study (Diniz, 2008) (Table 3, No. 63) was excluded because only two children were compared; one study (Dew-Hughes and Blandford, 1998) (Table 3, No. 103) because no clear distinction was made between students with and without Down syndrome; and three studies (Gheradini, 2000; Schramm, 1974; Vianello and Lanfranchi, 2009) (Table 2, No. 39 and 53; Table 3, No. 61) because a comparison was made with a reference group outside the study without precise information on this reference group. However, two of these studies (Gheradini, 2000; Schramm, 1974) (Table 2, No. 39 and 53) could be included as a non-comparative study containing information on acceptance by classmates. Out of the pool of potentially relevant non-comparative studies a majority of 53 were excluded because the study contained no information on social acceptance. Also excluded were 20 single case studies, 6 studies on attitudes of school children to children with Down syndrome in general (instead of own classmates with Down syndrome) and 5 mixed population studies in which no clear distinction was made between students with and without Down syndrome.

As can be seen in table 3, a few studies were excluded for more than one reason.

Table 1. Included comparative studies on the effect of (pre)school placement on students with Down syndrome

Study	Country	N	Longitudinal/ Cross-sectional	Age in yrs	Data sources	Effect* of regular placement				Also reported in:
						Self-help	Language	Academics	Social	
1. Matthews (2009)	Ireland	65	C	12-19	Parent questionnaires	-	-	-	0 Having a close friend; AR Having a group of friends; AR (s) Having a friend without a disability; AR (s) Speaking to many people during social activities; AR (s) Number of social activities in clubs	
2. Turner et al. (2008)	UK	71	L	6-14 at t1; 11-19 at t2; 18-26 at t3	Teacher and parent questionnaire	-	-	AR (s)	-	
3. Vervat (2008)	NL	43	C	8-13	Parent scale and PSPCSA by the child	AR (s)	AR (s)	-	0 Self-esteem	
4. Dijkxhoorn et al. (2007)	NL	504	C	4-18	Vineland screener (parents)	AR (s)	-	-	-	
5. de Graaf (2007a;b)	NL	121	C	5-13	Parent questionnaire	AR (s)	AR (s)	AR (s)	AR (s) Friendships at home; AR (s) Having a friend without a disability; AS (s) Having a friend with a disability; AS (s) Equal friendships at school	
6. Buckley, Bird, Sacks et al. (2006)	UK	46 (now) + 90 (older cohort)	C	11-20	Parent questionnaire; Vineland; Conners rating scale	0	AR (s)	AR (s)	0 Socialisation Vineland; AR (s) Behavioural problems + social mature behaviour; AS (s) Interpersonal relationships	Buckley et al. (2002a;b)

Table 1. (Continued)

Study	Country	N	Longitudinal/ Cross-sectional	Age in yrs	Data sources	Effect* of regular placement				Also reported in:
						Self-help	Language	Academics	Social	
7. McMahon (2003)	Ireland	38	C	6-16	Parent questionnaire (adaptation of Vineland)	-	0	-	0 Socialisation and behavioural problems	
8. Cuckle and Wilson (2002)	UK	14	C	12-18	Interviews with students, classmates, parents and teachers	-	-	-	AS Friends at school	
9. Freeman and Kasari (2002)	USA	27	C	5-11	Direct observation	-	-	-	AS (s) Quality of interaction during play date with chosen friend	
10. Bochner et al. (2001)	Australia	30	C	18-36	Normative tests on reading and language	-	AR (s)	AR (s) Reading	-	
11. Laws et al. (2000)	UK	44	C	7-14	Normative tests on reading, language and memory.	-	AR (s)	AR (s) Reading	-	
12. Begley (1999)	UK	64	C	8-16	PSPCSA (child)	-	-	-	0 Self-esteem	
13. Gould (1998)	UK	24	C	11-18	PSPCSA (child)	-	-	-	0 Self-esteem	
14. Beadman (1997)	UK	22	C	4-12	IQ-test; ratings of behaviour by teachers	-	AR	-	0 Behavioural problems	
15. Bronson et al. (1997)	USA	115 (38 DS)	C	5	IQ; systematic observation	-	-	AR (s) Task-oriented behaviour	AR(s) Using adequate social strategies	
16. Bochner and Pieterse (1996)	Australia	87	C	13-20	Parent questionnaire	0	-	AR	AR Independent behaviour	

Study	Country	N	Longitudinal/ Cross-sectional	Age in yrs	Data sources	Effect* of regular placement				Also reported in:
						Self-help	Language	Academics	Social	
17. Yadarola (1996)	Argentina	54 (C); and 10 (L)	C (54); and L (10)	5-10 (C); 5-10 (L) at t1, 9-14 at t4	Teacher and parent questionnaires and interviews; classroom and recess observations	-	AR	AR	AR Number of interactions; Positive interactions; Child's satisfaction with school and with school tasks	Yadarola (1998)
18. Laws et al. (1995)	UK	14	L	4-10 at t1; 8-14 at t2	Normative tests on reading, language and non-verbal IQ	-	AR (s)	AR (s)	-	
19. Hauser-Cran and Bronson (1993)	USA	148 (49 DS)	C	3	IQ; systematic observation	-	-	AR (s) Task-oriented behaviour	AR (s) Using adequate social strategies	
20. Philps (1992)	UK	30	C	5-11	Normative tests on language and IQ, teacher questionnaire	-	0	AR (s)	AR (s) Number of interactions at playtime; AS (s) Number of language initiations to peers	
21. Fewell and Oelwein (1990)	USA	135 (58 DS)	L	3-10 (6 mths between t1 and t2)	Criterium-referenced test	0	AS (s)	-	0 Social development	
22. Sloper, Cunningham et al. (1990)	UK	181	C	6-14	Teacher and parent questionnaire	AR (s)	-	AR (s)	-	
23. Sloper, Turner et al. (1990)	UK	181	C	6-14	Teacher and parent questionnaire	-	-	-	0 Social contacts after school	
24. Casey et al. (1988)	UK	36	L	4-10	Normative tests on language, math, MA	-	AR (s)	AR (s)	-	
25. Lorenz et al. (1985)	UK	115	C	5-7	IQ; MA; teacher questionnaire on reading;			AR (s)		
26. Sinson and Wetherick (1981)	UK	7	L	2-5	Direct observation of interactions with peers				AS Eye contact and social interactions	

*: 0= no difference; AR= Advantage for Regular students; AS= Advantage for Special students; (s)= statistically significant difference ($p<0.05$).

Table 2. Included descriptive studies on the social acceptance of regularly placed students with Down syndrome

Study	Country	N	Longitudinal/ Cross-sectional	Age in yrs	Data sources	Also reported in:
27. Dolva (2009)	Norway	6	L	10	Participant observation and interviews with staff and child with Down syndrome	Dolva et al. (2010)
28. Guralnick et al. (2009)	USA	81 (27 DS)	C	4-7	Parent questionnaires on social network; normative tests on language and IQ; parent rating scales for behaviour; Vineland	
29. Klompas (2008)	South Africa	3	C	8-12	Interviews with parents and teachers	
30. Castro (2007)	Colombia	60	C	3-16	Questionnaires for school staff, parents, counsellors and persons with Down syndrome	
31. Wilson (2007)	UK	35 (18 DS)	C	5-16	interviews with staff; rating of self-esteem (competence and social acceptance) with the PSPCSA (child)	
32. de Graaf (2006)	NL	15	L	4-13	Interviews with parents, regular teachers and counsellors from special education; narrative observations; parent and staff questionnaires.	
33. Hamilton (2005)	Australia	30 (10 DS)	L	4-5	Observations of interactions	
34. Solorzano Arriaga (2005)	Guatemala	5	C	5-12	Observations of interactions, interviews with teachers and parents	
35. Commodari and Pirrone (2004)	Italy	119	C	11-15	Sociometric ratings by classmates	
36. Shevlin et al. (2003)	Ireland	3	C	13-15	Interviews with parents (looking in retrospective at integration in primary school)	
37. Rietveld (2002)	New Zealand	4 (2 DS)	L	4-5	Observations of interactions; interviews with staff and parents.	Rietveld (2008)
38. de Graaf (2001)	NL	12	C	5-13	Interviews with parents and teachers; narrative observations.	de Graaf (2002)
39. Gheradini (2000)	Italy	385	C	6-15	Questionnaires for teachers and head teachers	
40. Quail (2000)	UK	7	C	11-18	Observations of interactions	
41. de Graaf (1998)	NL	18 (16 DS)	C	5-16	Interviews with parents	
42. Kliewer (1998a;b)	USA	10	L	2-10	Participant observation and interviews with staff	
43. Scheepstra (1998)	NL	23	C	6-8	Observations of interactions; ratings of social acceptance by teachers; sociometric rating by classmates.	Scheepstra et al. (1999); Pijl and Scheepstra (1996;1998)

Study	Country	N	Longitudinal/ Cross-sectional	Age in yrs	Data sources	Also reported in:
44. Laws et al. (1996)	UK	16	C	8-11	Sociometric ratings by classmates	
45. Wolpert (1996)	USA	250	C	4-20	Parent and teacher questionnaire	
46. Petley (1993)	UK	10	C	6-8	Interviews with parents and headmasters	
47. Rietveld (1989)	New Zealand	6	L	6-7 at t1; 9-11 at t2	Observations of interactions	
48. Rietveld (1986)	New Zealand	8	C	6-7	Observations of interactions; teacher questionnaire.	
49.Hudson and Clunies-Ross (1984)	Australia	15 (11 DS)	C	5-8	Sociometric ratings by classmates; ratings by parents and teachers; observations of interactions	
50. Pieterse and Center (1984)	Australia	8	C	7-9	Sociometric ratings by classmates; teacher ratings.	
51. Knox (1983)	Australia	6	L	3-7	Observations of interactions	
52. Rogers-Warren et al. (1980).	USA	4	C	5-6	Observations of interactions	
53. Schramm (1974)	USA	2	C	8-10	Interviews with parents and teachers	

Table 3. Studies excluded from the review

Study	Country	Reason* for exclusion
54. Odluyurt and Batu (2010)	Turkey	C
55. Tanti Burlò (2010)	Malta	C
56. Wendelborg and Tøssebro (2010)	Norway	F
57. Beltrame et al. (2009)	Brazil	E
58. Doherty and Egan (2009)	Ireland	C
59. Gannon and McGilloway (2009)	Ireland	D
60. Hooton and Westaway (2009)	UK	C
61. Vianello and Lanfranchi (2009)	Italy	B
62. Silva (2009)	Brazil	C;E
63. Diniz (2008)	Brazil	A
64. Keenan (2008)	Ireland	C
65. Koulousia (2008)	Greek	C
66. Morrison (2008)	South Africa	E
67. Muniz (2008)	Brazil	C
68. Neto and Silva (2008)	Brazil	C
69. Sirlopu et al. (2008)	Chile	D
70. Gannon and McGilloway (2007)	Ireland	D
71. Sioutis (2007)	Greek	C
72. Alvarez and Ramirez (2006)	Colombia	C
73. Casale-Giannola andWilson Kamens (2006)	USA	E
74. Johnson (2006)	UK	C
75. Engelbrecht et al. (2005)	South Africa	C
76. van Hove et al. (2005)	Belgium	F
77. Kenny et al. (2005)	Ireland	C
78. Rynders (2005)	USA	C
79. Solórzano Arriaga (2005)	Guatemala	C
80. Butler and Hodge (2004)	USA	E
81. Down's Syndrome Association (2004)	UK	C
82. Felice (2004)	Argentina	C
83. Fox et al. (2004)	UK	C
84. Campbell et al. (2003)	Australia	C
85. Eloff and Kriel (2003)	South Africa	C
86. McCormick et al. (2003)	USA	C;E
87. Clarke (2002)	Ireland	D
88. Kostelnik et al. (2002)	USA	E
89. Mori (2002)	Argentina	C
90. Newmark (2002)	South Africa	C
91. Engelbrecht et al. (2001)	South Africa	C
92. Gaad (2001)	United Arabic Emirates	C
93. Vianello and Moalli (2001)	Italy	D
94. Wang et al. (2001)	USA	C
95. Wolpert (2001a;b)	USA	C
96. Egan-McGann (2000)	Ireland	C
97. Gash et al. (2000)	France, Ireland, Portugal, Spain	D
98. Muthukrishna et al. (2000)	South Africa	C
99. Freeman et al. (1999)	USA	C
100. Kasari et al. (1999)	USA	C

Study	Country	Reason* for exclusion
101. Martins (1999)	Brazil	C
102. Velez et al. (1999)	Colombia	C
103. Dew-Hughes and Blandford (1998)	UK	F
104. Giangreco et al. (1998)	USA	E
105. Cuckle (1997)	UK	C
106. Engelbrecht et al. (1997)	South Africa	C
107. Rao (1997)	Hong Kong	C
108. Sader (1997)	South Africa	C
109. Wybranski (1997)	USA	C
110. Cheney and Demchak (1996)	USA	E
111. Lorenz (1996)	UK	C
112. Muthukrishna (1996)	South Africa	C
113. Scheepstra et al. (1996)	NL	C
114. Cormany (1994)	USA	C;F
115. Passaro (1994)	USA	C;E
116. Fox and Hanline (1993)	USA	C
117. Barringer (1992)	USA	E
118. Elias (1991)	USA	E
119. West and Cummins (1990)	USA	C
120. Center (1989)	Australia	F
121. Allen (1987)	UK	E
122. Fredericks et al. (1987)	USA	C
123. Bookbinder (1986)	UK	E
124. Budgell (1986)	UK	C
125. Elias et al. (1986)	USA	E
126. Fredericks (1986)	USA	E
127. Humphreys (1984)	USA	C
128. Centre for Studies on Integration in Education (1983)	UK	C;E
129. Elias et al. (1983)	USA	E
130. Bruni (1982)	USA	E
131. Cooke (1982)	USA	C;E
132. Hayes et al. (1981)	Australia	C;F
133. Gorelick and Brown (1974)	USA	C

*: A= Comparative study with only two children; B= Study in which a comparison is made with a reference group outside the study, without presenting adequate and precise information on this reference group; C= Non-comparative study, however without information on the extent of social acceptance of the students with Down syndrome by their peers; D= Study exploring peer attitudes to children with Down syndrome in general, not to real classmates with Down syndrome; E= Non-comparative single case study; F= Study in which only a minority has Down syndrome and in presenting the results no distinction is made between students with and without Down syndrome.

Selective Placement or Differential Stimulation?

To understand differences in development between specially and regularly placed students with Down syndrome, two processes should be disentangled: selective placement versus differential stimulation. Do children with Down syndrome acquire more academics in regular education because the children with more potential have a higher chance to be in regular education? Or, do they learn more academics because regular education is more stimulating? For practical and ethical reasons, it is impossible to conduct at random trials of placements. So, researchers usually use one of two different approaches, or a combination of both, to differentiate between the effects of selective placement versus differential stimulation. The first approach is looking at the effect of placement on a dependent variable (for instance academics) while statistically correcting for the possible effects of other relevant child and/or family variables (for instance mental age of the child or parental educational level). The second approach is looking at natural experiments in which school placement was not determined by child characteristics but by geographical area and/or generation.

Self-Help Skills

Seven studies (Table 1, No. 3-6, 16, 21, 22) could be found in which a comparison between placements was made with regard to self-help skills of children with Down syndrome. In four of these studies, notably an English study from Sloper, Cunningham et al. (1990) (No. 22) and three Dutch studies of respectively Vervat (2008), Dijkxhoorn, Oudheusden, and Berckelaer-Onnes (2007) and de Graaf (2007b) (No. 3, 4, and 5), a difference was found, i.e., the regularly placed students were more advanced. However, this could very well be the result of selective placement. In research conducted by Sloper, Cunningham et al. (No. 22), the difference between regularly and specially placed students in self-help skills disappeared after controlling for differences in mental age. Furthermore, in both a study of Buckley, Bird, Sacks et al. (2006) (also reported in Buckley, Bird, Sacks and Archer, 2002a;b) (No. 6) and Fewell and Oelwein (No. 21), no significant differences in self-help skills were found. It is important to note that the study of Buckley, Bird, Sacks et al. (2006) can be considered to be a natural experiment, in which according to Buckley and colleagues school placement was not determined by child characteristics but by differences in educational policy in different geographical areas. Finally, in a research of Bochner and Pieterse (1996) (No. 16) no differences were found in daily living skills between a relatively recent cohort of Australian teenagers with Down syndrome, of which more than half had been in regular education for most of their school career, and older more segregated cohorts.

There is no evidence that regular placement of children with Down syndrome leads to a better development of self-help skills, but neither there is any evidence for an advantage of special school placement in this regard. It might very well be the case that parents of children with Down syndrome have more influence on the development of their child's self-help skills than school placement.

Language

In twelve studies (Table 1, No. 3, 5-7, 10, 11, 14, 17, 18, 20, 21, 24), researchers evaluated language development. In only one of these, notably Fewell and Oelwein (1990) (No. 21), a difference was found in favour of special placement. However, Fewell and Oelwein themselves don't attribute the higher gains in expressive language of children with Down syndrome in special preschool and school settings to the placement itself, but to the fact that in the special preschool settings an effective program for early intervention was carried out in a more intensive way. In this program strong emphasis was put on stimulating speech and language skills, among others by using whole sight word reading for language enrichment. Apparently, it is not where the children are educated, but more what is educated in a particular setting and with what intensity. However, an even more important issue in the context of this review is a methodological characteristic of this study. The more segregated children were younger than the more integrated and the researchers did not statistically control for age. Whereas development in Down syndrome is not necessarily linear, this might flaw the comparison between settings.

In a study by Philps (1992) (No. 20), no differences in formal language measures (Reynells Developmental Language scales) were found between regularly and specially placed children with Down syndrome. In another study, by McMahon (2003) (No. 7), results indicated that school placement had no statistically significant effect on the development of communicative adaptive skills in children with Down syndrome. However, the advantage in communication skills of the regularly placed children nearly reached statistical significance ($p=0.052$), despite the fact that the specially placed children on average were two years older. Yet, according to McMahon, this result of no difference remained the same when age and gender differences were taken into consideration. However, for the huge differences in calendar age of the specially and regularly placed children, correction was made in a rather rough way, only by dividing the group in children under and above twelve years of age. Moreover, in this specific comparison between specially and regularly placed children, McMahon only looked at actual school placement, without taking the child's school history into account. Consequently, students who might have been in mainstream education for almost their entire school career, and only lately had been transferred to a special school, were considered to be specially placed students. Interestingly, McMahon made two other comparisons. Both children who had previously attended mainstream preschool and children who had previously attended mainstream school displayed significantly higher communication results ($p<0.01$) than did children who had never attended another school (ten out of eleven of the students in this latter group had started their school career in special education).

In all the other nine studies (Table 1, No. 3, 5, 6 , 10, 11, 14, 17, 18, 24), differences were found in favour of regular classroom placement. In five of these (No. 3, 5, 10, 14, 17) no attempt was made to disentangle the effect (on language skills) of selective placement from the effect of differences in stimulation. However, three other studies, of Buckley, Bird, Sacks et al. (2006) (No. 6), Laws et al. (2000) (No. 11) and Casey et al. (1988) (No. 24), can be considered to be natural experiments. School placement was not determined by child characteristics but by differences in educational policy in different geographical areas. This implies that we may attribute the differences in language development in these three studies to differential stimulation between settings. Moreover, in the Casey et al. study (No. 24), regularly placed children showed higher gains in receptive language, also after controlling for age, mental age and initial scores on language and academics. In the research of Laws et al. (2000) (No. 11), regularly placed students had higher scores on diverse measures for language skills, also after controlling for differences in receptive vocabulary (which normally correlates with mental age). And, in the study of Buckley, Bird, Sacks et al. (2006) (No. 6), regularly placed students had much higher scores on speech and language skills than specially placed students, but no differences were found in daily living skills nor in the overall socialisation score on the Vineland Adaptive Behaviour Scales. Finally, in a small longitudinal study of Laws, Buckley, Bird, MacDonald, and Broadley (1995) (No. 18) on 14 students with Down syndrome, at age 4-10 years no differences between settings in language, memory and non-verbal cognitive development were found. At age 8-14 however, 6 out of 7 regularly placed children and 1 out of 7 specially placed children had some reading abilities (on the BAS). 'Readers' were more advanced on measures for language development at age 8-14. There were no differences in non-verbal abilities.

On basis of the available recent studies we may conclude that regular classroom placement has considerable advantages for the language development of children with Down syndrome. Especially in the study of Buckley, Bird, Sacks et al. (2006) (No. 6), a large effect was found. The regularly placed teenagers in this particular study had on average scores of 2 years and 6 months higher on speech and language skills on the Vineland (and moreover, in a parent questionnaire on average their speech was rated as more articulate).

Academics

Fourteen studies (Table 1, No. 2, 5, 6, 10, 11, 15-20, 22, 24, 25) related to academic skills. In each of these, regularly placed students with Down syndrome had better academic skills than their specially placed counterparts. This also applies after correcting for differences in non-academic cognitive functioning, as demonstrated in the studies of Turner, Alborz, and Gayle (2008) (No. 2), de Graaf (2007b) (No. 5), Buckley, Bird, Sacks et al. (2006) (No. 6), Bronson, Hauser-Cran, and Warfield (1997) (No. 15), Hauser-Cran and Bronson (1993) (No. 19), Philps (1992) (no. 20), Sloper, Cunningham et al. (1990) (no. 22), Casey et al. (1988) (No.24) and Lorenz, Sloper, and Cunningham (1985) (No. 25). In addition, three studies of Buckley, Bird, Sacks et al. (2006) (No. 6), Laws et al. (2000) (No. 11) and Casey et al. (1988) (No. 24) can be considered to be natural experiments in which school placement was not determined by child characteristics but by geographical area. In one more study, of Laws et al. (1995) (No. 18), no differences between settings in language, memory and non-verbal cognitive development were found early in the school career of the children with Down syndrome at age 4-10 years. At age 8-14 however, 6 out of 7 regularly placed children and 1 out of 7 specially placed children had some reading abilities (on the BAS). Finally, making a comparison with older almost totally segregated generations of students with Down syndrome, both the study of Buckley, Bird, Sacks et al. (2006) (No. 6) and Bochner and Pieterse (1996) (No. 16) show a considerable advantage in academic skills in the more recent cohorts of teenagers with Down syndrome, of which many had been in regular education for most of their school career. The Buckley, Bird, Sacks et al. (2006) study demonstrates that this generational difference is totally explained by the higher level of academic skills of the students with Down syndrome in regular educational settings in the more recent cohort.

We may conclude that the differences in academic skills in favour of regularly placed children with Down syndrome cannot be the result merely of selective placement in regular schools of the more able children. It also is a direct result of more stimulation of academic development in regular settings. Turner et al. (2008) (No. 2) and Sloper, Cunningham et al. (1990) (No. 22) found a modest beneficial effect on academic skills of mainstream attendance. In the studies of de Graaf (2007b) (No. 5), Buckley, Bird, Sacks et al. (2006) (No. 6), Laws et al. (2000) (No. 11), Yadarola (1996) (No. 17) and Laws et al. (1995) (No. 18), even considerable advantages of regular placement on the development of academic skills were demonstrated. Moreover, the beneficial effects of regular classroom placement have been proven for diverse aspects of school learning, notably task-oriented behaviour (Bronson et al., 1997 (Table 1, No. 15); Hauser-Cran and Bronson, 1993 (No. 19); Yadarola, 1996 (No. 17)), reading (Bochner, Outhred, and Pieterse, 2001 (No. 10); Buckley, Bird, Sacks et al., 2006 (No. 6); de Graaf, 2007b (No. 5); Laws et al., 2000 (No. 11); Laws et al., 1995 (No.18); Lorenz et al., 1985 (No. 25); Philps, 1992 (No. 20); Sloper, Cunningham et al., 1990 (No. 22); Turner et al., 2008 (No. 2); Yadarola, 1996 (No. 17)), writing (Buckley, Bird, Sacks et al., 2006 (No. 6); de Graaf, 2007b (No. 5); Philps, 1992 (No. 20); Sloper, Cunningham et al., 1990 (No. 22); Turner et al., 2008 (No.2); Yadarola, 1996 (No.17)) and math (Buckley, Bird, Sacks et al., 2006 (No. 6); Casey et al., 1988 (No. 24); de Graaf, 2007b (No. 5); Sloper, Cunningham et al., 1990 (No. 22); Turner et al., 2008 (No. 2); Yadarola, 1996 (No. 17)).

Social Aspects

As regards social aspects, in eighteen studies (Table 1, No. 1, 3, 5-9, 12-17, 19-21, 23, 26), a comparison was made between regular and special placement. The results show a mixed image.

In ten studies (Table 1, No. 1, 3, 6, 7, 12, 13, 14, 17, 21, 23), no significant differences between settings were found in social aspects, notably regarding social skills (McMahon, 2003) (No. 7), gains in social skills (Fewell and Oelwein, 1990) (No. 21), social network (Sloper, Turner, Knussen, and Cunningham, 1990) (No. 23), self-esteem (Begley, 1999 (No. 12); Gould, 1998 (No. 13); Vervat, 2008 (No. 3)), behavioural problems (Beadman, 1997 (No. 14); Buckley, Bird, Sacks et al., 2006 (No. 6); McMahon, 2003 (No. 7)) and having a close friend (Matthews, 2009) (No. 1). Additionally, in a study by Yadarola (1996) (No. 17), teachers in special and regular primary education rated their students with Down syndrome more or less similarly on social behaviour, peer interactions and acceptance, with the exception that, in comparison with peers in special schools, peers in regular schools were more highly rated on being protective towards their classmate with Down syndrome and on helping the child with school work, a finding by Yadarola interpreted as a sign of a supportive peer group in regular schools. In the study of Buckley, Bird, Sacks et al. (2006) (No. 6), specially placed students had significantly more behavioural problems on the Vineland Adaptive Behaviour Scales, but not on the Conners Rating Scales. Furthermore, it is important to note that most mainstreamed students with Down syndrome in the research of Sloper, Turner et al. (1990) (No. 23) were not in their neighbourhood school. This might explain why in this particular study regular placement didn't lead to more opportunities for playing with other children at home. As regards self-esteem, Glenn and Cunningham (2001) state that for persons with Down syndrome the validity of the measure used, the Pictorial Scale of Perceived Competence and Social Acceptance for Young Children, is doubtful. Too many of them rate themselves positively by choosing the most positive answer instead of really considering their own situation.

Eight studies (Table 1, No. 1, 5, 6, 15, 16, 17, 19, 20) demonstrated advantages of regular classroom placement. As regards social mature behaviour, Bochner and Pieterse (1996) (No. 16), using parent questionnaires, reported more social mature behaviour in their recent cohort of Australian teenagers with Down syndrome, of which more than half had been in regular education for most of their school career, than in older more segregated cohorts. Buckley, Bird, Sacks et al. (2006) (No. 6) reported a similar difference between generations. In addition, according to the parent questionnaire conducted on the more recent cohort in the study of Buckley and colleagues, the regularly placed teenagers with Down syndrome had more social mature behaviour than their specially placed counterparts (and less behavioural problems according to the Vineland). As has been stated, in this particular study placement was not determined by child characteristics but by differences in educational policy in different geographical areas. Furthermore, Bronson et al. (1997) (No. 15) and Hauser-Cran and Bronson (1993) (No. 19) demonstrated in a study of respectively 115 and 148 preschoolers with developmental disabilities (one third Down syndrome) that social mature behaviour (using more adequate social strategies), and task-oriented behaviour as well, was higher in more integrated settings, also after controlling for cognitive ability of the child and demographic and socio-economic characteristics of the family. Philps (1992) (No. 20) reported that the children with Down syndrome in mainstream took part in more interactions at playtime than those in schools for children with Moderate Learning Difficulties (MLD). Based on observations, Yadarola (1996) (No. 17) reported that during recess in the special schools the social climate tended to be either one of poor control of aggression, with not enough surveillance by staff, resulting in aggressive behaviour in some children and fear in others, or one of overprotection, resulting in children mostly interacting with staff instead of peers. In the regular schools, social climate was more balanced, promoting positive interaction with peers. Although some of the children with Down syndrome in regular schools encountered overprotection by teachers and peers (actually more often than in special schools), this didn't result in an overall overprotective climate during recess. As regards social network, on the basis of parent questionnaires, de Graaf (2007a) (No. 5) reported that regularly placed students with Down syndrome had more opportunities for playing with other children (not siblings) at home than their specially placed counterparts, also after correcting for differences in non-academic skills. Furthermore, students in regular education more often had a friend without a disability, a result not only found by de Graaf (2007a) but by Matthews (2009) (No. 1) as well. In addition, in Matthew's study regularly placed children more often participated in social activities in clubs and they were more frequently rated as speaking to many people during social activities.

In six studies (Table 1, No. 5, 6, 8, 9, 20, 26) a disadvantage of regular placement was reported. Four of these studies (No. 5, 6, 8, 17) made use of questionnaires and/or interviews with students, parents and/or teachers. We firstly discuss the findings from these studies. In comparison with specially placed teenagers, regularly placed teenagers with Down syndrome had less opportunities for friendships with other peers with a developmental disability and as a result for the development of equal close friendships at school (Buckley, Bird, Sacks et al.,

2006 (No. 6); Cuckle and Wilson, 2002 (No. 8)). De Graaf (2007a) (No. 5) confirmed this finding for children with Down syndrome in primary education. However, it is important to note that, although specially placed students with Down syndrome had more opportunities for developing equal close friendships at school, these friendships seldom continued into the home situation (Cuckle and Wilson, 2002; de Graaf, 2007a), partly as a consequence of limited transport skills of the children and teenagers involved. Barriers to participation in leisure activities, according to a study of Matthews (2009) (No. 1) were not having a friend to go with, followed by not knowing how to do activities, not having a place nearby and not having a way to get there. Indeed, many parents of teenagers with Down syndrome (Bochner and Pieterse, 1996 (No. 16); Buckley, Bird, Sacks et al., 2006 (No. 6); Cuckle and Wilson, 2002 (No. 8)) worry about the relative social isolation of their child. Consequently, Cuckle and Wilson conclude that parents of children with Down syndrome should continue helping organize the social life of their child, long after the child has reached an age at which for children without Down syndrome there no longer is a need to do such.

Three other studies (Table 1, No. 9, 20, 26) made use of direct observation of interactions. Sinson and Wetherick (1981) (No. 26) videotaped the interactions of seven children with Down syndrome in a special preschool setting and of three of these children in a regular playgroup. In the regular playgroup situation, the children without a disability, on first encounter, attempted to make eye-contact with the children with Down syndrome. During this encounter, the children without a disability became increasingly uneasy, having no success in sustaining mutual gaze. Observations of the three children with Down syndrome in their regular playgroups over a period of two years indicated that the other children eventually gave up and the children with Down syndrome became isolates. In the special setting, however, the children with Down syndrome interacted with each other in much the same way as the children without a disability in the regular playgroup. There was ample evidence of mutual gaze as a precursor of verbal and play interactions. Sinson and Wetherick, clearly conceptualising from the deficit model of disability, suggest that children with Down syndrome have a deficit, demonstrated by untypical mutual gaze behaviour, directly leading to their isolation from typical peers. However, in another related article, Sinson and Wetherick (1986) state that interactions of the children with Down syndrome with their mothers, and with their siblings as well, were not affected. They conclude that the prevailing method of simply introducing a child with Down syndrome into an unknown preschool peer group unlikely will lead to successful social integration. On the other hand, according to Sinson and Wetherick (1986), prolonged contacts with peers in a family play situation with mothers present, from an early age, probably support social integration. The value of the study by Sinson and Wetherick is in demonstrating that social inclusion is not reached by mere placement without any support. However, Sinson and Wetherick fail to consider the possibility that not only mothers in a family play situation, but educators in preschool settings as well, have an opportunity to model and foster social interactions between children with and without disabilities.

Freeman and Kasari (2002) (No. 9) studied the interactions of dyads of children in play-date situations. Participants were 54 children, 27 with Down syndrome and 27 who were their chosen friend (the parents were asked to bring a friend of the child with Down syndrome to the play-date). In comparison with children with Down syndrome from regular education classes, children with Down syndrome from special education classes were rated as more cohesive and coordinated in their quality of play and they spent more time in simple social play and less time in solitary, parallel or parallel aware play. The children from general and special education were not different on demographic and developmental characteristics. This suggests that during these organised play-dates the interactions of the children with Down syndrome with their chosen friend more often displayed the characteristics of a close friendship (instead of just being playmates) if the child with Down syndrome was from a special school.

As regards another qualitative aspect of interaction, Philps (1992) (No. 20) conclude that, although children with Down syndrome in mainstream took part in more interactions at playtime than those in MLD, when initiations of language were examined it was clear that all the children initiated more language when they were in the company of younger or less able peers. This kind of setting occurred more often in MLD schools than in mainstream, except where children played with younger children at playtime in mainstream settings. In order to stimulate more language initiations, Philps suggest that children with Down syndrome in regular education should deliberately be put in the role of leader on occasions, instead of being in a situation with more able peers all the time.

Non-Comparative Studies on Social Aspects

In addition to these comparative studies, in 27 other non-comparative studies (Table 2), the acceptance of regularly placed children with Down syndrome by their classmates was evaluated.

Sixteen of these studies (Table 2, No. 29, 30-32, 35, 36, 38, 39, 41, 43-45, 46, 49, 50, 53) evaluated acceptance by exploring the perceptions of classmates, parents, teachers and/or head teachers. Regularly placed children with Down syndrome are generally fairly well accepted by their peers, according to sociometric evaluations (Commodari and Pirrone, 2004 (Table 2, No. 35); Hudson and Clunies-Ross, 1984 (No. 49); Laws et al., 1996 (No. 44); Pieterse and Center, 1984 (No. 50); Scheepstra, 1998 (No. 43)) and questionnaires and/or interviews with parents, teachers and/or head teachers (Castro, 2007 (Table 2, No. 30); Gheradini, 2000 (No. 39); de Graaf, 1998 (No. 41), 2001 (No. 38), 2006 (No. 32); Hudson and Clunies-Ross, 1984 (No. 49); Klompas, 2008 (No. 29); Petley, 1993 (No. 46); Pieterse and Center, 1984 (No. 50); Scheepstra, 1998 (No. 43); Schramm, 1974 (No. 53); Shevlin, Walsh, Kenny, McNeela, and Molloy, 2003 (No. 36); Wilson, 2007 (No. 31); Wolpert, 1996 (No. 45)). In addition, the study of Wilson (2007) (No. 31) shows that children with Down syndrome in regular education positively evaluate their acceptance by peers, rating themselves higher than children with general developmental delay (not Down syndrome), and even higher than children without a disability.

However, whereas the children in four of the sociometric studies had a generally high level of peer acceptance, in the research of Scheepstra (No. 35), though only one child had a 'rejected' sociometric status, 12 out of 23 children (6-8 years of age) had a 'neglected' status. Children with a 'neglected' status are not actively disliked or avoided, but they are less often chosen as preferred playmate. Furthermore, while in the study of Laws et al. (1996) (No. 44), considering all six sociometric questions, most children with Down syndrome (8-11 years of age) had an average social status, they were definitively less often seen as 'best' friend by classmates. It is important to realize that sociometric results are dependent upon the exact content of the questions posed to children. From the sociometric study of Commodari and

Pirrone (No. 35) in Italian middle school, it appears that children with Down syndrome frequently would be invited to birthday parties by classmates, however were more likely to be excluded from activities involving a certain degree of ability, like doing schoolwork together or joining a sport team. Commodari and Pirrone conclude that in middle school students without disabilities accept classmates with Down syndrome, but in practice during many school and leisure related activities prefer to socialise with peers without disabilities.

In addition, parents interviewed by de Graaf (1998) (No. 41) reported that in regular school their child's contacts with classmates gradually changed from interactions based on equality in preschool to interactions that resemble a relationship with an older sibling towards the end of primary school. Both headmasters and parents in the study of Petley (1993) (No. 46) recognized the risk of relationships in which the child is 'mollycoddled' by classmates and not treated with equal respect. According to Wilson (2007) (No. 31), school staff interviewed sometimes overestimated the degree of acceptance of children with Down syndrome by their peers, not recognizing the difference between supportive relationships and equal status relationships. Parents and teacher in the study of Klompas (2008) (No. 29) reported that the three children with Down syndrome (8-12 years of age) in this study, although well accepted, lacked reciprocal friendships. Therefore, parents of, particularly older, regularly placed students with Down syndrome should be advised to organize contacts also with peers with a disability, supplementary to the relations their child has with children without a disability. However, one could make a strong case to do just as well in the case of specially placed students, because the parents report that their special friendships with classmates are often confined to only the school hours (de Graaf, 2007a; Cuckle and Wilson, 2002) (Table 1 No. 5 and 8). Moreover, from the study of de Graaf (2007a) it is clear that specially placed children also have a larger chance of not having any friendship at all with children without a disability.

In another study (Guralnick, Connor, and Johnson, 2009) (Table 2, No. 28), acceptance by preschool peers was approached in a more indirect way, not through directly informing about peer acceptance in preschool, but asking mothers about the home-based social network of their child. In comparison with two control groups, a calendar age matched group and a mental age matched group, children with Down syndrome had a significantly smaller network size (mother-identified regular playmates, maximum of 5 children) and a less high frequency with which playmates played in mother's home in the previous 3 months. In addition, a smaller percentage of identified playmates were currently in the school program of the child with Down syndrome. Only around one third of the children with Down syndrome knew any of their playmates as a result of being classmates, whereas this was the case for nearly two thirds and over three quarters of the MA-match and CA-match groups, respectively. Guralnick et al. suggest that a major contributor to these peer social network limitations may well be underlying peer-related social competence difficulties in children with Down syndrome.

In twelve studies (Table 2, No. 27, 33, 34, 37, 38, 42, 43, 47, 48, 49, 51, 52), researchers explored acceptance by peers making (also) use of direct observation of interactions. An early study is by Rogers-Warren, Ruggles, Peterson, and Cooper (1980) (No. 52). In a mainstreamed preschool setting with 60 percent children with a disability, four 5-6 years old children with Down syndrome and four typically developing contrast children were observed. The children with Down syndrome and the contrast children preferred the same play areas in the classroom and in the playground. The children with Down syndrome had a higher frequency of solitary play, the contrast children a higher frequency of parallel play. The children with Down syndrome tended to play more often with other children with a disability. The contrast children more often chose to play with peers without a disability. However, interaction between children with Down syndrome and children without a disability occurred in all activities.

Eleven studies (Table 2, No. 27, 33, 34, 37, 38, 42, 43, 47, 48, 49, 51) focussed on individual inclusion with only one child with Down syndrome being included in a regular preschool or school. As we have seen in the previous section, Sinson and Wetherick (1981) (Table 1, No. 26) suggest that children with Down syndrome over time almost inevitably will become more and more isolated in regular preschool and school settings. In a recent study by Hamilton (2005) (Table 2, No. 33) a more or less similar result was found. Compared with typically developing preschool children, children with Down syndrome in inclusive preschool displayed infrequent interactions with peers. While typically developing children increased their peer interactions during activities expected to promote peer interaction (for instance during playtime), the interactive engagement of the children with Down syndrome was unrelated to the characteristics of class activities. Longitudinally, over a period of two years, in contrast to the typically developing children, children with Down syndrome did not increase their rates of peer interaction. Informal observations revealed that teachers in this study didn't use effective strategies to enhance peer interaction in children with Down syndrome. Hamilton concludes that their implicit strategy of only prompting children to be in physical proximity to other children, without actually encouraging interactive engagement, should be supplemented by explicit teaching strategies, like designing activities that require group interaction among children and adults, prompting and reinforcing peer interaction, modelling interactive behaviour, and interpreting for the typically developing children the meaning of the social behaviour of the children with disabilities.

Some other studies contradict the suggestion of Sinson and Wetherick that children with Down syndrome over time almost inevitably will become more and more isolated in regular preschool and school settings. Knox (1983) (Table 2, No. 51) observed six children with Down syndrome (3-7 years of age, two in preschool and four in school). During the 9 week observation period, the children with Down syndrome increased both their use of language to initiate interactions with peers as well as their responding to verbal directions by peers. Hudson and Clunies-Ross (1984) (No. 49) found no differences in the rates of positive and negative peer interactions of 11 regularly placed school children with Down syndrome (6-8 years of age) in comparison with 11 randomly selected contrast children from their classrooms. However, the children with Down syndrome initiated fewer interactions with peers than the typically developing contrast children.

This last result, less initiations to peers, was found in three other studies as well (Quail, 2000 (Table 2, No. 40); Rietveld, 1986 (No. 48); Scheepstra, 1998 (No. 43)). Quail (2000) (No. 40) carried out a study of the social interactions of 7 teenagers with Down syndrome in mainstream secondary schools in comparison to peers from the same classrooms. There were no differences in overall time spent in interaction. However, the other person more often initiated the interactions of the teenagers with Down syndrome and they had more interactions with adults than with peers. Topics in the conversations of the teenagers with Down syndrome were likely to be work related rather than socially related. In this specific study, these patterns of interactions might partly be the result of the fact that the teenagers were being supported by a Learning Support Assistant for most of the day. However, comparing 23 children with Down syndrome (6-8 years of age), in the Dutch situation with only a limited amount of support a week (8 hours), with 23 low achieving contrast children from the same classrooms, Scheepstra (1998) (No. 43) (also reported in Scheepstra, Nakken, and Pijl, 1999; and in Pijl and Scheepstra, 1996) observed fewer interactions with peers, more interactions with the teacher, fewer initiations to peers, and more work related and less socially related interactions by the children with Down syndrome. Finally, Rietveld (1986) (No. 48) compared eight children with Down syndrome (6-7 years of age) with eight low achieving contrast children from their classrooms. There were no differences in the rates of social play. However, the children with Down syndrome initiated fewer interactions to peers. Interestingly, Rietveld (1989) (No. 47) carried out a follow-up study of six of these children with Down syndrome, now 9-11 years of age. In comparison with the first study, she observed in the children with Down syndrome more initiations to peers and from peers and less interactions with the teacher. Also, a higher percentage of time was spent in social play and the interactions were more complex and varied with more advanced language. Remarkably, the difference between the children with Down syndrome and the contrast children in the rate of initiating to peers had disappeared.

Four studies (Dolva, 2009 (Table 2, No. 27); Kliewer, 1998a;b (No. 42); Rietveld, 2002 (No. 37); Solorzano Arriaga, 2005 (No. 34)) made use of a more qualitative interpretative methodology. By in depth exploration of small samples, this kind of research can cast light upon the complexities of the inclusion process and the role of the cultural and educational context. Dolva (2009) (No. 27) studied six students with Down syndrome in inclusive schools, in the classroom and in the playground, collecting data through field observations and interviewing the child and the staff. The children with Down syndrome were well accepted by peers. Dolva observed examples of equal interactions in which all children participating understood the activity and could interact on a rather equal basis. However more often she observed unequal interactions, with limited understanding of the activity by the child with Down syndrome and/or involving tasks that at first were too difficult. In these situations peers spontaneously divided tasks in a complementary way, or they adjusted tasks or own behaviour or created other tasks in order to include the pupil with Down syndrome without losing the original meaning of the activity. Class staff members were found to experience interaction between students with Down syndrome and peers challenging, but still possible because of peers' acceptance. Staff applied different strategies to facilitate peer interaction, for example: organising academic activities and group work; purposefully selecting tasks and pairing children to work together; educating peers to behave supportively; providing individual support to help the child to understand activities and scaffolding participation. Solorzano Arriga (2005) (No. 34) studied the social development of five girls with Down syndrome in two Guatemalan (non-special) private schools. She concludes that social integration in the schools should be supported by making curricular adaptations and that families can play an important role in preparing their child with Down syndrome for school integration by systematically stimulating the child's development form an early age onwards. Kliewer (1998a,b) (No. 42) explored the (pre)school literacy experience of 10 students with Down syndrome. Two broad definitions of literacy were uncovered. The first regarded reading as confirming to a hierarchy of sub-skills. In these classrooms students with Down syndrome were separated from citizenship in the classrooms' literate communities. The second regarded reading as the construction of shared meaning in specific contexts. In these classrooms, students with Down syndrome were valued as symbolic beings and engaged literacy as a communication tool. As a result, during the school day, there were more opportunities for participation and for demonstrating friendship. According to Kliewer, the first definition of literacy seems to be more in line with a deficit model of disability, interpreting difference as deficit, the second with the social model. Both Kliewer (No. 42) and Rietveld (2002) (No. 37) (also reported in Rietveld, 2008) highlight the consequences of the way staff conceptualises disability. Rietveld carried out an in depth study of the experiences of four boys, two with Down syndrome and two typically developing, during their transition from preschool to school. The children were observed using continuous narrative recordings during all aspects of the curriculum in preschool and in school. Teachers, parents and peers were interviewed. Inclusion or exclusion turned out not to be within-child characteristics, but largely dependent on the context, both for the children with and without Down syndrome, demonstrated by the huge changes in the extent of inclusion between the preschool and school setting for two of the boys. In one of the schools, staff modelled that it was acceptable to exclude when 'deviant' children did not fit the existing implicit or explicit classroom rules. Also, teacher and teacher-aide positively reinforced children when they assigned the child with Down syndrome an inferior status role, for instance the role of 'the object of a caregiver'. By contrast, in another school the teacher openly interpreted the likely intent of any unconventional behaviour in a positive and valuing manner. Furthermore, she included activities that highlighted the competencies and interests of her student with Down syndrome in a way that made the overall class culture more inclusive for many more children. The teacher and teacher-aide recognised and interrupted demeaning inclusion, e.g. excessive hugging, picking up. The staff scaffolded children to re-frame any problems they interpreted within a deficit framework to one that focussed on context. Successful outcomes of inclusion were associated with schools embracing at all levels a social model of disability, that focuses on the context and sees disability as a part of, not distinct from that context.

CONCLUSION

In the period 1970-2010, the context of regular placement of pupils with Down syndrome has changed from integration, to mainstreaming, to inclusive education. As a result, the way children with Down syndrome are supported in regular schools may have shifted over time. Furthermore, the organisation and curriculum of special schools for children with Down syndrome may have changed. In addition, also differences between countries in these regards may and probably will exist. The evidence found in this 40-year period of research is limited to twelve different countries, or regarding only comparative studies to only six countries. As regard the beneficial effect of regular schooling on, particularly, academic and language development of children with Down syndrome, despite these cultural, organisational and educational differences between these countries, the findings of the studies in this review converge to a considerable extent. This suggests that, even with these differences, there are sufficient similarities in the organisation and curriculum of special schools for children with Down syndrome and in the support of regularly placed pupils with Down syndrome over time and between different countries - at least between the countries from which the studies are derived - to make a review of research a sensible enterprise. For instance, starting in the mid 1980s and in many different countries, regular schooling for pupils with Down syndrome means education in a regular classroom with individual support to some extent. Furthermore, one could argue that special schools for children with Down syndrome, worldwide, will tend to focus more on practical and social skill acquisition, and as a result probably be liable to de-emphasise academic skill acquisition.

From our review it can be concluded that regular placement of students with Down syndrome, i.e., education in a regular classroom with individual support to some extent, yields a better development of language and academic skills, even after the effect of selective placement has been taken into account. In some of the studies a modest beneficial effect could be proven, in other studies even considerable advantages of regular placement on the development of language and academic skills were demonstrated. As regards self-help skills, under the same condition, there seem to be no differences.

As regards social functioning, the results show a mixed image. For some social aspects (social network, behaviour, self-competence) no differences at all or small positive differences for regularly placed children were found. However, studies also highlight that for many children with Down syndrome, mere placement in a regular setting without any support might not be enough. Social interactions between children with and without Down syndrome oftentimes need to be modelled and fostered. In addition, to get around the tendency of many children with Down syndrome in regular settings to respond to initiations of others rather than initiate to peers themselves, it is also advised to put children with Down syndrome sometimes in a leading position. Two qualitative studies suggest that in schools embracing a social model of disability, rather than a deficit model, the necessary social support to facilitate both more interactions, as well as more reciprocity and equal status in these interactions, seems to be more likely available. Furthermore, although regularly placed children are generally fairly well accepted by their peers, they are less often seen as 'best' friend. In addition, regularly placed teenagers with Down syndrome have fewer opportunities for friendships with other peers with a developmental disability and, as a result, for the development of equal close friendships at school. Therefore, parents of, particularly older, regularly placed students with Down syndrome should be advised to organize contacts also with peers with a disability, supplementary to the relations their child has with children without a disability. However, one could make a strong case to do just as well in the case of specially placed students, because the parents report that their special friendships with classmates are often confined to only the school hours.

It is important to mention the fact that in research on the effect of school placement, IQ and/or mental age usually are conceptualised as independent variables, i.e., control variables in determining the effect of placement on development, and not as dependent variables. However, it is conceivable that intelligence, aside of being a factor in determining placement, itself also might be directly influenced by school experience. In that case, cross-sectional studies in which IQ and/or mental age are used as control variables might underestimate the positive effect of regular placement on academic and language development. In a study by Rao (1997) on 6 preschoolers with Down syndrome, all attending a centre-based educational intervention program for three mornings a week, 3 of these children also attended regular preschools, two days a week. The researchers compared the children in the Program Only group with those in the Program Plus Preschool group. At the start of the study, both the mother's educational level and the child's IQ, chronological age and mental age were similar in both groups. A year later, the IQ of all 3 children in the Program Only group had decreased (on average minus 6 points, range from minus 3 to minus 10). The change in IQ scores for children in the Program Plus Preschool group was less dramatic (on average plus 1 point, range from minus 1 to plus 2). This study does not demonstrate the superiority of regular preschool placement in comparison to special preschool placement, because the 3 children not in regular preschool actually stayed at home on the days they didn't visit the special educational centre. However, acknowledging that the sample size was very small, the study does suggest that the development of IQ might be influenced by preschool experience. In earlier decades, debate was fierce on the question whether children with Down syndrome institutionalised at an early age had lower IQ's than home-reared children as a result of institutionalisation or as a result of being 'constitutionally inferior in intelligence' (Carr, 1988). Carr (1988), in her review of four decades of studies on IQ of children with Down syndrome, concludes that the evidence offers very little support for the 'constitutional inferiority' hypothesis. The alternative hypothesis, that the difference between home- and institution-reared children is due to environmental influence, according to Carr, not only accounts more simply for the facts, but is also in accord with the effects on these children of environmental enrichment. This debate might be relevant to the present discourse on the effects of school placement. Firstly, we encounter the same issue of disentangling the effect of selective placement from the effect of differences in stimulation between settings. Secondly, the former debate shows that IQ should be considered as a dependent variable, a variable that is influenced by experience and thus possibly by school experience as well. In this regard, Vianello and Moalli (2001) (Table 3, No. 93) and Gheradini (2000) (Table 2, No. 39) make the interesting suggestion that school aged children with Down syndrome in the Italian educational system, as a result of inclusion, might have a higher mean IQ than usually found in international research (according to Vianello and Moalli with an average IQ of 45 for Italian subjects and less than 40 for others); however, in both articles this hypothesis is not supported by a thorough analysis of different international studies. Nevertheless, although methodologically very complex, the suggestion of comparing the cognitive development of children with Down syndrome in different countries, contrasting inclusive educational systems with more segregated systems, could be an avenue for future research. Alternatively, in countries where some children with Down syndrome go to regular schools and others to special schools, another avenue could be following-up representative groups of children with Down syndrome throughout their school career, not only repeatedly measuring their skills, but also their mental age at different chronological ages.

Apart from studying the potential effect of an inclusive school system on mean total IQ in children with Down syndrome, Vianello and Lanfranchi (2001) (Table 3, No. 61) suggest that it might be interesting to look at their profile of development. Vianello and Lanfranchi present two Italian studies, one on mental age and academic development in school-aged children with Down syndrome, and one on intellectual level and scores on daily activities and socialization in adolescents with Down syndrome. Both studies show that in many cases the age-equivalent of specific skills (academics, especially reading; daily living skills; social skills) is higher than the age-equivalent of mental development. According to Vianello and Lanfranchi, this 'surplus' is not usually found in international research and might be the result of inclusive education. However, this hypothesis is not supported by a thorough analysis of different international studies. Vianello and Lanfranchi only refer to research of Zigler and colleagues over a period of 40 years, stating that persons with a intellectual disability oftentimes demonstrate a 'deficit' in skills in respect to their mental age, as a result of having less motivation to work and having less self-esteem. However, firstly, this research was not

Down syndrome specific, secondly the personality and motivational characteristics of the persons in Zigler's research might have been influenced by institutionalisation. Vianello and Lanfranchi conclude that, in their article they have formulated more questions than generated answers. The affirmation that 'surplus' in academic and social performance seems to be greater where academic inclusion of persons with a disability is more widespread, is still a hypothesis to confirm. However, the study of Buckley, Bird, Sacks et al. (2006) (Table 1, No. 6) is in agreement with the hypothesis that education can influence the developmental profile. In a paper accompanying their study, titled 'Evidence that we can change the profile from a study of inclusive education', Buckley, Bird, and Sacks (2006) conclude that the Down syndrome specific profile – a profile of communication weaknesses relative to social and daily living skills – can be changed. It is not an inevitable outcome of having Down syndrome, as this profile is seen in teenagers in special education settings, but not in teenagers in inclusive education.

As regards the beneficial effect of regular classroom placement on language development, different mechanisms might play a role. Firstly, the regular classroom seems to be a richer language environment, with more challenging language being used by teaching staff, as a small explorative UK study of Dew-Hughes and Blandford (1998) (Table 3, No. 103) suggests. Based on observations, these researchers reported that classroom language was less sophisticated in the special schools. Language was simple and direct, allowing for comprehension by the less fluent members of the class. In mainstream schools, language was more challenging. Secondly, peers in a regular classroom will most certainly be behavioural examples using more complex language than those in a special classroom. Yadarola (1996) (Table 1, No. 17) reported that the children with Down syndrome in Argentinean special education were surrounded by peers who mostly used signs and/or very short-spoken utterances for communication, whereas peers in regular school used more complex language. As a result, children with Down syndrome in regular schools were invited and encouraged to understand and use more complex language themselves, whereas in special schools the children with Down syndrome adapted to the special-school-specific communication culture of signing and using simple phrases. Finally, more emphasis on academic skills in regular classrooms, especially on reading activities, might foster understanding and use of more complex language.

Only a very limited amount of research has been done on the reasons why regular school placement, i.e., education in a regular classroom with individual support to some extent, leads to better academic skills in children with Down syndrome. Lorenz et al. (1985) (Table 1, No. 25) start with denying the contention that most children with Down syndrome in special schools are not taught to read, as in their birth cohort of children with Down syndrome, 38 percent of teachers from the schools for children with Severe Learning Difficulties reported to teach reading to the children with Down syndrome and another 33 percent reported to teach prereading skills. However, Lorenz et al. state that their data suggest that when children with Down syndrome at a three- to five-year-level are matched for mental age, those in regular schools score highest on a checklist of reading skills and those in schools for children with Severe Learning Difficulties score lowest. Lorenz et al. suggest that an explanation might be that in the special school curriculum there is a greater emphasis on the development of prerequisites, including fine motor and language skills, before reading is taught, resulting in an unnecessary postponement from prereading to reading activities. The point at which teachers move pupils from prereading to reading activities may be important in the acquisition of reading skills. Furthermore, according to Lorenz et al., studies on teachers' aims suggest that teachers in special schools generally place much less emphasis on reading skills compared to teachers in regular primary schools. These findings are corroborated by both the cross-sectional and longitudinal research of Yadarola (1996) (Table 1, No. 17). In this study, teachers from special schools placed less emphasis on academic skill development. The transfer from teaching prerequisites to teaching reading, writing and math, was postponed. Most children with Down syndrome at age 9 and 10 in the Argentinean special schools had not yet been exposed to any instruction at all in reading, writing or math. Teachers in special schools strongly believed that before starting instruction in academics, certain prerequisite skills should have been developed. In sharp contrast, in regular schools, the onset of teaching these skills to children with Down syndrome usually was around the age of 5 or 6. In addition, two small explorative UK studies, of Beadman (1997) (Table 1, No. 14) and Dew-Hughes and Blandford (1998) (Table 3, No. 103), report that in special schools most teacher time was devoted to the least able children in the classroom. Effective teaching time was greatly reduced as a result of time spent on transport, physical care regimes, therapies and slower-moving members of the group. The teaching staff in special schools had lower expectations for academic achievement of their students with Down syndrome than teaching staff in mainstream schools. In special schools, classrooms were generally poorly equipped with reading materials and schemes. Furthermore, in regular school students with Down syndrome received more individual instruction time (in the UK half to fulltime personal assistance is quite common for mainstreamed children with Down syndrome). Also in the study of Philps (1992) it turned out that there was a significantly greater time spent in one to one work in mainstream than in the MLD schools studied. The Dutch research of de Graaf (2007b) confirms that in the Netherlands as well, although regular schools receive less money for extra assistance time for a student with Down syndrome than in the UK (in the Netherlands on average 4 to 8 hours extra teacher's time a week), students with Down syndrome still receive more individual instruction time in regular schools than in special schools. In addition, it was demonstrated that in regular schools children with Down syndrome on average spent one and a half to twice as much time on academic learning than in special schools.

Regular placement of students with Down syndrome has a beneficial effect on language and academic skills. However, it is important to note that merely regular placement is not enough. In the different studies regular placement means placement in a regular classroom with individual support to some extent, though varying between countries and regions and between individual students. Cunningham et al. (1998) and Cuckle (1997) demonstrate that in the nineties in the UK huge local differences existed in the extent to which students with Down syndrome started their school career in regular schools and in the extent to which they stayed in regular education to at least the end of primary education. These differences must be the result of varying local educational policies, probably with as a result differences in support of mainstreamed children with Down syndrome. So, an important question is what amount and kind of support is needed in order to make regular placement feasible for individual children with Down syndrome. In the Netherlands, according to de Graaf (2007a), approximately 56 percent of all children with Down syndrome from the years of birth 1993-2000 started their school career in regular education. Of children with Down syndrome who started their school career at a regular school slightly more than 40 percent continued in regular education for the entire elementary school period. This implies that transferring mainstreamed children with Down syndrome to special school during the elementary years is quite common. Again, this points at the question what amount and kind of support is needed to make regular placement feasible for individual children with Down syndrome. Only a very limited amount of research has been done on the question why the regular school career of some children with Down syndrome comes to an early end while others continue in regular education for the entire elementary school period. From the few (Dutch) systematic studies (de Graaf, 2006, 2007a; Poulisse, 2002; Scheepstra, 1998) it can be concluded that not only child characteristics but also school characteristics determine whether a child with Down syndrome will succeed or not in regular education. Some schools succeed in providing inclusive education for very challenging children. De Graaf (2006) investigated twenty cases of 'problematic integration situations'. These included 13 children with Down syndrome whose regular school placement was in danger at that moment, 4 children who were only recently placed in a special school after one or more years of regular education and 3 children who had just started anew at another regular school after being sent away from their first regular school. In each case, parents, teaching staff from the regular school, and counsellors from special education (who advise the regular schools) were interviewed about their perceptions of why a 'problematic integration situation' had emerged. In relation to school characteristics, according to parents and counsellors from special education, differences in staff attitude and school vision on inclusion, and according to parents, counsellors and regular teachers alike, the way teachers, assistants, counsellors and parents manage to work together determine whether a child will succeed or not at a certain regular school. However, this important issue deserves a more elaborate analysis which is beyond the scope of this review.

It is important to note that research on the effects of inclusive education on children without disabilities shows that regular classroom placement of students with disabilities has no negative impact on the development of classmates (Cole, Waldron, and Majd, 2004; Hunt and Goetz, 1997; Hollowood, Salisbury, Rainforth, and Palombaro, 1994; Manset and Semmel, 1997; McDonnell, Thorson, McQuivey, and Kiefer-O'Donnell, 1997; McIntosh, Vaughn, Schumm, Haager, and Lee, 1993; Saint-Laurent et al., 1998; Salend and Garrick Duhaney, 1999; Sharpe, York, and Knight, 1994; Staub, 2005; Staub and Peck, 1994/95; Waldron and Cole, 2000). According to some studies, it even brings them positive social advantages like more pro-social behaviour (Allodi, 2002; Hunt and Goetz, 1997; York, Vandercook, Macdonald, Heise-Neff, and Caughey, 1992), better moral development (Dumke and Mergenschröer, 1990; Staub, 2005; Staub and Peck, 1994/95; York et al., 1992) and a more accepting and less stereotyping attitude toward persons with disabilities (Fisher, 1999; Lehrer, 1983; Salend and Garrick Duhaney, 1999; Scheepstra, 1998; Staub, 2005; Staub and Peck, 1994/95; York et al., 1992) and towards persons from other minorities (Fisher, 1999; Kishi and Meyer, 1994; Staub, 2005; Staub and Peck, 1994/95).

In the light of the proven advantages of regular placement for the development of language and academics, more opportunities for inclusion of children with Down syndrome in regular classrooms should be created. Besides this, inside the special schools children with Down syndrome should receive more attention in regard to the teaching of academic skills. We acknowledge that children with Down syndrome are not a homogenous group. Turner and Alborz (2003) rightly argue that a small percentage of children with Down syndrome are unable to achieve any significant level of literacy or numeracy and that the needs of these children should not be overlooked by placing too much emphasis on academics. However, presently, for the vast majority of children with Down syndrome in special schools often too much of their learning potential in this area remains unused. Furthermore, social interactions with other children, at school and outside school hours, for many children with Down syndrome need to be organized and supported deliberately. This is independent of the school type the child is going to. Finally, embracing a social model of disability, rather than a deficit model, seems to be facilitative for social inclusion.

Acknowledgment

The authors wish to express their thanks to Emma de Moel, who corrected the English.

References

Allen, L. (1987). *Like other children.* London, England: Centre for Studies on Integration in Education. Retrieved from ERIC database. (ED292243).

Allodi, M. W. (2002). A two-level analysis of classroom climate in relation to social context, group composition, and organization of special support. *Learning Environments Research, 5*, 253-274.

Alvarez, O. H., and Ramírez, D. (2006). Experiencias e investigaciones sobre las TIC aplicadas a la atención de personas con necesidades educativas (in Spanish) [Experience with and research on ICT in supporting people with special educational needs]. *Revista Comunicación y Pedagogía, 213*, 56-63.

Baker, E. T., Wang, M. C., and Walberg, H. J. (1994/95). The effects of inclusion on learning. *Educational Leadership, 52*, 33-35.

Barringer, M. D. (1992). Mainstreaming. Wanted: Just one friend. Case History. *Learning, 21*(4), 29.

*Beadman, J. (1997). *An evaluation of educational placement for children with Down's syndrome in the South Devon Education Area.* Devon: Devon Psychological Service.

*Begley, A. (1999). The self-perceptions of pupils with Down syndrome in relation to their academic competence, physical competence and social acceptance. *International Journal of Disability, Development and Education, 46*(4), 515-530.

Beltrame, T. S., Herber, V., and Ribeiro, J. (2009). O Processo de socialização de uma criança com síndrome de Down no sistema regular de ensino (in Portuguese) [The process of socialization of a child with Down syndrome in the regular education system]. *Dynamis revista tecno-cíentíca, 2*(15), 38-43.

*Bochner, S., and Pieterse, M. (1996). Teenagers with Down syndrome in a time of changing policies and practices: Progress of students who were born between 1971 and 1978. *International Journal of Disability, Development and Education, 43*(1), 75-95.

*Bochner, S., Outhred, L., and Pieterse, M. (2001). A study of functional literacy skills in young adults with Down syndrome. *International Journal of Disability, Development and Education, 48*(1), 67-90.

Bookbinder, G. (1986). Professionals and wrong predictions. *British Journal of Special Education, 13*(1), 6-7.

*Bronson, M. B., Hauser-Cran, P., and Warfield, M. E. (1997). Classrooms matter: Relations between the classroom environment and the social and mastery behavior of five-year-old children with disabilities. *Journal of Applied Developmental Psychology*, *18*, 331-348.

Bruni, S. L. (1982). *The class and Faye: Social interaction of a handicapped child in an integrated first grade class*. Retrieved from ERIC database. (ED271905).

Buckley, S., and Bird, G. (2000). *Education for individuals with Down syndrome - An overview. Down Syndrome Issues and Information*. Portsmouth: The Down Syndrome Educational Trust.

*Buckley, S., Bird, G., Sacks, B., and Archer, T. (2002a). A comparison of mainstream and special education for teenagers with Down syndrome: Implications for parents and teachers. *Down Syndrome News and Update*, *2*(2), 46-54.

*Buckley, S., Bird, G., Sacks, B., and Archer, T. (2002b). The achievements of teenagers with Down Syndrome. *Down Syndrome News and Update*, *2*(3), 90-96.

Buckley, S., Bird, G., and Sacks, B. (2006). Evidence that we can change the profile from a study of inclusive education. *Down Syndrome Research and Practice*, *9*(3), 51-53.

*Buckley, S., Bird, G., Sacks, B., and Archer, T. (2006). A comparison of mainstream and special school education for teenagers with Down syndrome: Implications for parents and teachers. *Down Syndrome Research and Practice*, *9*(3), 54-67.

Budgell, P. (1986). Drifting towards segregation. *British Journal of Special Education*, *13*(3), 94-95.

Butler, R. S., and Hodge, S. R. (2004). Social inclusion of students with disabilities in middle school physical education classes. *Research in Middle Level Education Online*, *27*(1), 1-10.

Campbell, J., Gilmore, L., and Cuskelly, M. (2003). Changing student teachers' attitudes towards disability and inclusion. *Journal of Intellectual and Developmental Disability*, *28*(4), 369-379.

Carlberg, C., and Kavale, K. (1980). The efficacy of special versus regular class placement for exceptional children: a meta-analysis. *The Journal of Special Education*, *14*, 295-307.

Casale-Giannola, D., and Kamens, M. W. (2006). Inclusion at a university: experiences of a young woman with Down syndrome. *Mental Retardation*, 44(5), 344-352.

*Casey, W., Jones, D., Kugler, B., and Watkins, B. (1988). Integration of Down's syndrome children in the primary school: a longitudinal study of cognitive development and academic attainments. *The British Journal of Educational Psychology*, *58*(3), 279-286.

*Castro, C. A. B. (2007). *Revision del estado de la práctica de los procesos de inclusión educativa de los estudiantes con síndrome de Down en Colombia* (in Spanish) [*Review of the state of practice of the process of educational inclusion of students with Down syndrome in Colombia*]. Bogotá, Colombia: Corporacion Sindrome de Down-ASDOWN, Septiembre de 2007.

Center, Y. (1989). *The integration of children with disabilities into regular schools. A naturalistic study. Stage 2 report*. Macquarie University, Sydney. Retrieved from ERIC database. (ED323696).

Centre for Studies on Integration in Education. (1983). *Kirsty: The struggle for a place in an ordinary school. CSIE Fact Sheet*. London, England: Centre for Studies on Integration in Education. Retrieved from ERIC database. (ED298717).

Cheney, C., and Demchak, M. (1996). *Providing appropriate education in inclusive settings: A rural case study*. Retrieved from ERIC database. (ED394761).

Clarke, P. (2002). *Peer and teacher attitudes towards the integration/inclusion of children with Down's syndrome into mainstream primary schools* (master's thesis). University College Dublin, Dublin, Ireland.

Cole, C. M., Waldron, N., and Majd, M. (2004). Academic Progress of Students Across Inclusive and Traditional Settings. *Mental Retardation*, *42*(2), 136-145.

*Commodari, E., and Pirrone, C. (2004). L'integrazione scolastica del disabile: il punto di vista degli alunni (in Italian) [School integration of the disabled: the point of view of students]. *Annali della facoltà di Scienze della formazione Università degli studi di Catania*, *3*, 91-101.

Cooke, N. L., Heron, T. E., Heward, W. L., and Test, D. W. (1982). Integrating a Down's syndrome child in a classwide peer tutoring system: A case report. *Mental Retardation*, *20*(1), 22-25.

Cormany, E. (1994). *Enhancing services for toddlers with disabilities: A reverse mainstreaming inclusion approach.* Practicum Report, Nova South Eastern University. Retrieved from ERIC database. (ED373496).

Cuckle, P. (1997). The school placement of pupils with Down syndrome in England and Wales. *British Journal of Special Education*, *24*, 175-179.

*Cuckle, P., and Wilson, J. (2002). Social relationships and friendships among young people with Down's syndrome in secondary schools. *British Journal of Special Education*, *29*(2), 66-71.

Cunningham, C., Glenn, S., Lorenz, S., Cuckle, P., and Shepperdson, B. (1998). Trends and outcomes in educational placements for children with Down syndrome. *European Journal of Special Needs Education*, *13*(3), 225-237.

Dew-Hughes, D., and Blandford, S. (1998). The social development of children with severe learning difficulties. *Down Syndrome Research and Practice*, *6*(1), 1-18.

*Dijkxhoorn, Y. M., Oudheusden, M. A. G., and Berckelaer-Onnes, I. A. (2007). Het adaptief functioneren van kinderen met Downsyndroom (in Dutch) [Adaptive functioning of children with Down syndrome]. *Down + Up (Journal of the Dutch Down Syndrome Foundation)*, *78*, 41–44.

Diniz, A. C. D. (2008). *O processo de inclusão da criança com síndrome de Down em contextos escolares diferenciados* (in Portuguese) [*The process of inclusion of children with Down syndrome in different school contexts*] (master's thesis). UNIPAC, Barbacena, Brazil.

Doherty, U., and Egan, M. (2009). *The report on educational provision for pupils with Down syndrome in mainstream primary schools in Limerick City and County.* University of Limerick, Limerick, Ireland.

*Dolva, A-S. (2009). *Children with Down syndrome in mainstream schools: conditions influencing participation* (PhD thesis). Karolinska Institutet, Stockholm, Sweden.

*Dolva, A-S., Hemmingsson, H., Gustavsson, A., and Borell, L. (2010). Children with Down syndrome in mainstream schools: peer interaction in activities. *European Journal of Special Needs Education*, *25*(3), 283-294. doi:10.1080/08856257.2010.492941.

Down's Syndrome Association. (2004). *Access to Education. A report on the barriers to education for children with Down's syndrome*. Middlesex, UK: Down's Syndrome Association.

Dumke, D., and Mergenschröer, B. (1990). Sozial Kognitionen von Schüler in Integrationsklassen. *Psychologie in Erziehung und Unterricht*, *37*, 111-122.

Egan-McGann, M. (2000). *Students who have Down syndrome: A study of their school placement educational supports and parental evaluation of their provision* (PhD thesis). National University of Ireland, Maynooth, Kildare.

Elias, L. (1991). Jason goes to junior high. *Exceptional Parent, 21*(6), 20-21.

Elias, L., Francisco-Medvin, R., Schefer, B., and Tenn, E. (1986). Jason goes to first grade. *Exceptional Parent, 16*(5), 12-13.

Elias, L., Goble, G., Schefer, B., and Jaco, K. (1983). Jason goes to kindergarten. *Exceptional Parent, 13*(1), 55-58.

Eloff, I., and Kriel, T. (2003). Ouers se konstruering van hul rol by insluitende onderwys van hulle kind met Downsindroom (in African*)* [Construction of their role in inclusive education by parents of a child with Down's Syndrome]. *South African Journal of Education, 23*(4), 328–337.

Engelbrecht, P., Eloff, I., Newmark, R., and Kachelhoffer, A. (1997). The Down's syndrome projects in inclusive education. *South African Journal of Education, 17*(2), 81-84.

Engelbrecht, P., Oswald, M., Swart, E., Kitching, A., and Eloff, I. (2005). Parents' experience of their rights in the implementation of inclusive education in South Africa. *School Psychology International, 26*, 459-477.

Engelbrecht, P., Swart, E., and Eloff, I. (2001). Stress and coping skills of teachers with a learner with Down syndrome in inclusive classrooms. *South African Journal of Education, 21*(4), 256-260.

Felice, M. (2004). *La integración escolar y las adaptaciones curriculares en la escuela común* (in Spanish) [*School integration and curricular adaptations in the regular school*]. Memorias del V Congreso Virtual de Educación. http://saci.org.br/?modulo=akemiandparametro=13797 (retrieved 17-11-2011).

*Fewell, R. R., and Oelwein, P. L. (1990). The relationship between time in integrated environments and developmental gains in young children with special needs. *Topics in Early Childhood Special Education, 10*(2), 104-116.

Fisher, D. (1999). According to Their Peers: Inclusion as High School Students See It. *Mental Retardation, 37*(6), 458-467.

Fox, L., and Hanline, M. F. (1993). A preliminary evaluation of learning within developmentally appropriate early childhood settings. *Topics in Early Childhood Special Education, 13*(3), 308-327.

Fox, S., Farrell, P. and Davis, P. (2004). Factors associated with the effective inclusion of primary-aged pupils with Down's syndrome. *British Journal of Special Education, 31*(4), 184-190.

Fredericks, B. (1986). I helped my son into the mainstream. *Exceptional Parent, 16*(5), 14-17.

Fredericks, H. D, Mushlitz, J., and DeRoest, C. (1987). Integration of children with Down syndrome at the elementary school level: a pilot study. In S.M. Pueshel, C. Tingey, J.E. Rynders, A.C. Crocker and D.M. Crutcher (Eds.), *New perspectives on Down syndrome* (pp. 179-193). Baltimore: Paul H. Brookes.

*Freeman, S. F. N., and Kasari, C. (2002). Characteristics and qualities of the play dates of children with Down syndrome: Emerging or true friendships? *American Journal on Mental Retardation, 107*(1), 16–31.

Freeman, S. F., and Alkin, M. C. (2000). Academic and social attainments of children with mental retardation in general education and special education settings. *Remedial and Special Education, 21*, 3-18.

Freeman, S. F., Alkin, M. C., and Kasari, C. L. (1999). Satisfaction and desire for change in educational placement for children with Down syndrome: Perceptions of parents. *Remedial and Special Education*, *20*, 143–151.

Gaad, E. E. N. (2001). Educating children with Down's syndrome in the United Arab Emirates. *British Journal of Special Education*, *28*(4), 195-203.

Gannon, S., and McGilloway, S. (2007). Taking steps toward inclusion: Assessing the attitudes of schoolchildren toward their peers with Down Syndrome. *The Frontline of Learning Disability*, *71*, 12-14.

Gannon, S., and McGilloway, S. (2009). Children's attitudes toward their peers with Down syndrome in schools in rural Ireland: an exploratory study. *European Journal of Special Needs Education*, *24*(4), 455-464.

Gash, H., Guardia Gonzales, G., Pires, M., and Rault, C. (2000). Attitudes towards Down syndrome: A national comparative study: France, Ireland, Portugal, and Spain. Irish *Journal of Psychology*, *21*, 203-214.

*Gherardini, P. (2000). *Inclusion in the mainstream school*. Paper presented at the 7th World Down Syndrome Congress, 23-26 March, Sydney, Australia.

Glenn, S., and Cunningham, C. (2001). Evaluation of self by young people with Down syndrome. *International Journal of Disability, Development and Education*, *48*(2), 163-177.

*Gould, S. (1998). *Self concept in teenagers with Down syndrome* (master's thesis). University of Portsmouth, Portsmouth, UK.

*de Graaf, G. W. (1998). Students with a developmental disability, particularly Down syndrome, in regular education: parents' ideas and approaches. In E. A. B. de Graaf, A. Vermeer, H. S. A. Heymans and M. I. M. Schuurman (Eds.), *Down syndrome behind the dykes: Research in the Netherlands* (pp. 155–172). Amsterdam: VU University Press.

*de Graaf, G. W. (2001). Een wereld van verschil: Interventies gericht op de sociale integratie van basisschoolleerlingen met Downsyndroom (in Dutch) [A world of differences. Supporting the social inclusion of students with Down syndrome in mainstream primary education]. *Special with Down + Up (Journal of the Dutch Down Syndrome Foundation)*, *55*.

*de Graaf, G. W. (2002). Supporting the social integration of students with Down syndrome in mainstream education. *Down Syndrome News and Update*, *2*(2), 55-62.

*de Graaf, G. W. (2006). Over de grenzen van integratie. Een onderzoek in het kader van een interventieproject bij problematische onderwijsintegratie-situaties (in Dutch) [On the boundaries of integration. A study in the context of an intervention project directed at 'problematic educational integration situations']. *Special with Down + Up (Journal of the Dutch Down Syndrome Foundation)*, *74*.

*de Graaf, G. W. (2007a). Onderwijs aan leerlingen met Downsyndroom. Deel 1: kind-, school-, en ouderkenmerken (in Dutch) [Education of students with Down syndrome. Part 1: characteristics of child, school and parents]. *Down + Up (Journal of the Dutch Down Syndrome Foundation)*, *79*, 46–58.

*de Graaf, G. W. (2007b). Onderwijs aan leerlingen met Downsyndroom. Deel 2: het effect van schoolplaatsing op de ontwikkeling (in Dutch) [Education of students with Down syndrome. Part 2: the effect of school placement on development]. *Down + Up (Journal of the Dutch Down Syndrome Foundation)*, *80*, 37–54.

Giangreco, M. F., Whiteford, T., Whiteford, L., and Doyle, M. B. (1998). Planning for Andrew: The use of COACH and VISTA in an inclusive preschool program. *International Journal of Disability, Development and Education*, *45*(4), 375-396.

Gorelick, M. C., and Brown, P. A. (1974). *Preschools willing to integrate children with handicaps. Directory 1974.* California State University. Retrieved from ERIC database. (ED090745).

*Guralnick, M. J., Connor, R. T., and Johnson, L. C. (2009). Home-based peer social networks of young children with Down syndrome: A developmental perspective. *American Journal on Intellectual and Developmental Disabilities*, *114*(5), 340–355.

*Hamilton, D. (2005). An ecobehaviour analysis of interactive engagement of children with developmental disabilities with their peers in inclusive preschools. *International Journal of Disability Development and Education*, *52*(2), 121-137.

*Hauser-Cram, P., Bronson, M. B., and Upshur, C. C. (1993). The effects of the classroom environment on the social and mastery behavior of preschool children with disabilities. *Early childhood research quarterly*, *8*(4), 479-498.

Hayes, A., Steinberg, M., Cooksley, E., Jobling, A., Best, D., and Coulston, A. (1981). *Special preschools: monitoring a pilot project. Volume I: Project Report.* Fred and Eleanor Schonell Educational Research Centre, University of Queensland, Australia. Retrieved from ERIC database. (ED245808).

Hollowood, T. M., Salisbury, C. L., Rainforth, B., and Palombaro, M. M. (1994). Use of instructional time in classrooms serving students with and without severe disabilities. *Exceptional Children*, *61*(3), 242-253.

Hooton, J., and Westaway, A. (2009). The voice of the child with Down syndrome. *Down Syndrome Research and Practice*, *12*(3), 25-29.

van Hove, G., Mortier, K., and de Schauwer, E. (2005). *Onderzoek inclusief onderwijs* (in Dutch) [*Research on inclusive education*]. Ghent University, Section Orthopedagogics, Ministry of the Flemish Community, Ghent, Belgium.

*Hudson, A., and Clunies-Ross, G. (1984). A study of the integration of children with intellectual handicaps into regular schools. *Australia and New Zealand Journal of Developmental Disablilities*, *10*(3), 165-177.

Humphreys, B. (1984). *Evaluating and modifying the educational expectations which parents and educators perceive as appropriate for Down's syndrome children.* Retrieved from ERIC database. (ED252297).

Hunt, P., and Goetz, L. (1997). Research on inclusive educational programs, practices, and outcomes for students with severe disabilities. *The Journal of Special Education*, *31*(1), 3-29.

Johnson, D. (2006). Listening to the views of those involved in the inclusion of pupils with Down's syndrome into mainstream schools. *Support for Learning*, *21*(1), 24-29.

Kasari, C., Freeman, S. F. N., Bauminger, N., and Alkin, M. C. (1999). Parental perspectives on inclusion: effects of autism and Down syndrome. *Journal of Autism and Developmental Disorders*, *29*(4), 297-305.

Keenan, M. (2008). *Speech language and communication difficulties in children with* Down syndrome*: An investigation of teachers' perceptions and practices in mainstream primary schools and special schools/classes in the counties of Longford Westmeath Offaly* (master's thesis). St Angela's College, Sligo, Ireland.

Kenny, M., Shevlin, M., Walsh, P. N., and McNeela, E. (2005). Accessing mainstream: examining the struggle for parents of children who have learning difficulties. *Journal of Research in Special Educational Needs, 5*(1), 11-19.

Kishi, G. S., and Meyer, L. H. (1994). What children report and remember: A six-year follow up of the effects of social contact between peers with and without disabilities. *Journal of the Association for Persons with Severe Handicaps, 19*, 277-289.

*Kliewer, C. (1998a). *Schooling children with Down Syndrome. Toward an understanding of possibility.* New York/ London: Teachers College Press.

*Kliewer, C. (1998b). Citizenship in the literate community: An ethnography of children with Down Syndrome and the written word. *Exceptional Children, 64*(2), 167-180.

*Klompas, M. S. (2008). *Inclusive education of primary school aged children with Down Syndrome in Gauteng Province, South Africa* (PhD thesis). University of Witwatersrand, South Africa. http://wiredspace.wits.ac.za/ handle/10539/4953 (retrieved 21-11-2011).

*Knox, M. (1983). Changes in the frequency of language use by Down's syndrome children interacting with non-retarded peer. *Education and Training of the Mentally Retarded, 18*(3), 185-190.

Kostelnik, M. J., Onaga, E., Rohde, B., and Whiren, A. (2002). *Children with special needs: lessons for early childhood professionals.* Early Childhood Education Series. Teachers College Press, Williston. Retrieved from ERIC database. (ED463899)

Koulosia, G. (2008). *Το προφίλ των μαθηματικών γνώσεων και δεξιοτήτων των παιδιών-ατόμων με σύνδρομο Down: μία ερευνητική προσέγγιση* (in Greek) [*The profile of mathematical knowledge and skills in children with Down syndrome: a research approach*] (master's thesis). University of Macedonia, Greece.

*Laws, G., Byrne, A., and Buckley, S. (2000). Language and memory development in children with Down syndrome at mainstream schools and special schools: a comparison. *Educational Psychology, 20*(4), 447-458.

*Laws, G., Buckley, S., Bird, G., MacDonald, J., and Broadley, I. (1995). The influence of reading instruction on language and memory development in children with Down's syndrome. *Down Syndrome: Research and Practice*, *3*(2), 59-64.

*Laws, G., Taylor, M., Bennie, S., and Buckley, S. (1996). Classroom behaviour, language competence, and the acceptance of children with Down syndrome by their mainstream peers. *Down Syndrome Research and Practice, 4*(3), 100-112.

Lehrer, A. (1983). The effects of mainstreaming on stereotypic conception of the handicapped. *The Journal of Educational Research, 77*(2), 94-97.

Lindsay, G. (2007). Educational psychology and the effectiveness of inclusive education-mainstreaming. *The British Journal of Educational Psychology, 77*, 1-24.

Lorenz, S. (1995). The placement of pupils with Down's syndrome: A survey of one Northern LEA. *British Journal of Special Education, 22*(1), 16-19.

*Lorenz, S., Sloper, T., and Cunningham, C. (1985). Reading and Down's syndrome. *British Journal of Special Education, 12*(2), 65-67.

Manset, G., and Semmel, M. I. (1997). Are inclusive programs for students with mild disabilities effective? A comparative review of model programs. *The Journal of Special Education, 31*(2), 155-180.

Martins, L. A. R. (1999). Integração escolar do portador da síndrome de Down: um estudo sobre a percepção dos educadores (in Portuguese) [School integration of people with

Down Syndrome: A study on the perception of educators]. *Revista Brasileira de Educação Especial, 5*, 73-85.

*Matthews, L., and Caffrey, S. (2009). *A comparative study of the social activities, participation and friendships of Irish adolescents with Down syndrome attending mainstream and special schools.* Poster presented at the 10th Word Down Syndrome Congress, 20-22 August, Dublin, Ireland.

McCormick, L., Wong, M., and Yogi, L. (2003). Individualization in the inclusive preschool: a planning process. *Childhood Education, 79*(4), 212-217.

McDonnell, J., Thorson, N., McQuivey, C., and Kiefer-O'Donnell, R. (1997). Academic engaged time of students with low-incidence disabilities in general education classes. *Mental Retardation, 35*(1), 18-26.

McIntosh, R., Vaughn, S., Schumm, J. S., Haager, D., and Lee, O. (1993). Observations of students with learning disabilities in general education classrooms. *Exceptional Children, 60*(3), 249-261.

*McMahon D. (2003). *A comparative analysis of the social and communicative adaptive behaviour of children with Down's syndrome who attend mainstream and special schools* (master's thesis). University College Dublin, Dublin.

Mori, R. (2002). *¿Es posible la integración de los niños con necesidades educativas especiales (NEE), a través del uso de la tecnología como herramienta didáctica*? (in Spanish) [*Is it possible to integrate children with special educational needs (SEN), through the use of technology as a teaching tool?*]. http://www. horizonteweb.com/magazine/ investigacionNEE.rtf (retrieved 17-11-2011).

Morrison, L. (2008). *The High School experience of a learner with Down syndrome: a case study* (master's thesis). Stellenbosch University, South Africa.

Muthukrishna, N., Farman, R., and Sader, S. (1996). *The inclusion of children with Down syndrome: A South African experience.* Springfield, VA: Resources in Education.

Muthukrishna, N., Sader, S., and Farman, R. (2000). The inclusion of children with Down syndrome in ordinary schools: The South African experience. *International Journal of Special Education, 15*(1), 86-95.

Muniz, H. S. (2008). *O Educando com síndrome de Down: um estudo sobre a relação entre pais e profissionais na escola regular* (in Portuguese) [*The student with Down syndrome: a study on the relationship between parents and professionals in regular school*] (master's thesis). Universidade Federal do Rio Grande do Norte, Brazil.

Neto, A. R. M., and Silva, T. J. C. (2008). Inclusão educacional de alunos com Síndrome de Down (in Portuguese) [Educational inclusion of students with Down syndrome]. *Revista Pensar a Prática, 11*(2), 105-113.

Newmark, R. (2002). *Inclusive education for learners with Down syndrome: The role of the educational psychologist* (PhD thesis). University of Stellenbosch, South Africa.

Odluyurt, S., and Batu, E. S. (2010). Teaching inclusion preparation skills to children with developmental disabilities. *Educational Sciences: Theory and Practice, 10*(3), 1563-1572.

Passaro, P. D. (1994). Using mastery learning to facilitate the full inclusion of students with the most intense educational needs within rural schools. *Rural Special Education Quarterly, 13*(3), 31-39.

*Petley, K. (1994). An investigation into the experiences of parents and head teachers involved in the integration of primary aged children with Down's syndrome into mainstream school. *Down Syndrome Research and Practice, 2*(3), 91-96.

*Philps, C. (1992). *Comparative study of the language development of children with Down syndrome in mainstream and special schools* (master's thesis). University of Wolverhampton, UK.

*Pieterse, M., and Center, Y. (1984). The integration of eight Down's syndrome children into regular schools. *Australia and New Zealand Journal of Developmental Disabilities, 10*(1), 11-20.

*Pijl, S. J. P., and Scheepstra, A. J. M. (1996). Being apart or a part of the group: The position of pupils with Down's syndrome in Dutch regular schools. *European Journal of Special Needs Education, 11*(3), 311-320.

*Pijl, S. J. P., and Scheepstra, A. J. M. (1998). Pioneering in integration: Pupils with Down syndrome attending regular education. In E. A. B. de Graaf, A. Vermeer, H. S. A. Heymans and M. I. M. Schuurman (Eds.), *Down syndrome behind the dykes: Research in the Netherlands* (pp. 173–182). Amsterdam: VU University Press.

Poulisse, N. (2002). *Een wankel evenwicht. De integratie van kinderen met een verstandelijke handicap in het reguliere basisonderwijs* (in Dutch) [*A delicate balance. The integration of children with intellectual disabilities in regular education*]. Nijmegen: ITS.

*Quail, S. (2000). *The social interactions of teenagers with Down syndrome and their non-disabled peers in inclusive school placements* (BSc Thesis). University of Portsmouth, UK.

Rao, N. (1997). Facilitating development in preschoolers with Down's syndrome in Hong Kong: A controlled evaluation study of the effects of centre-based intervention and regular preschool attendance. *British Journal of Special Education, 24*(4), 180-185.

*Rietveld, C. M. (1986). The adjustment to school of eight children with Down's syndrome from an early intervention program. *Australia and New Zealand Journal of Developmental Disabilities, 12*(39), 159-175.

*Rietveld, C. M. (1989). *The development of children with Down's syndrome who are mainstreamed in the Christchurch area: a study of six children observed in 1984/5 and in 1988.* Paper presented at the New Zealand Education Association Conference 'Reaching Beyond', at the Christchurch College of Education, 30, 31 August, 1 September 1989. Christchurch.

*Rietveld, C. M. (2002). *The transition from preschool to school for children with Down syndrome: a challenge to regular education?* (PhD thesis). University of Canterbury.

*Rietveld, C. M. (2008). Contextual factors affecting inclusion during children's transitions from preschool to school. *Australian Journal of Early Childhood, 33*(3), 1-9.

*Rogers-Warren, A., Ruggles T.R., Peterson N.L. and Cooper A.Y. (1980). *Playing and learning together: patterns of social interaction in handicapped and nonhandicapped children.* University of Kansas, Early Childhood Institute, Lawrence. Retrieved from ERIC database. (ED231104).

Ruijs, N. M., and Peetsma, T. T. D. (2009). Effects of inclusion on students with and without special educational needs reviewed. *Educational Research Review, 4*, 67-79. doi:10.1016/j.edurev.2009.02.002.

Rynders, J. (2005). Down syndrome: literacy and socialization in school. *Focus on Exceptional Children, 38*(1), 1-12.

Sader, S. B. (1997). *Educating children with Down Syndrome in regular schools: The KwaZulu-Natal experience* (master's thesis). University of Natal, South Africa.

Saint-Laurent, L., Dionne, J., Giasson, J., Royer, E., Simard, C., and Piérard, B. (1998). Academic achievement effects of an in-class service model on students with and without disabilities. *Exceptional Children, 64*(2), 239-253.

Salend, S. J., and Garrick Duhaney, L. M. (1999). The impact of inclusion on students with and without disabilities and their educators. *Remedial and special education, 20*(2), 114-126.

*Scheepstra, A. J. M. (1998). *Leerlingen met Down's syndroom in de basisschool* (in Dutch) [*Students with Down syndrome in regular primary school*] (PhD Thesis). University of Groningen. Groningen: Stichting Kinderstudies.

*Scheepstra, A. J. M., Nakken, H., and Pijl, S. J. (1999). Contacts with classmates: The social position of pupils with Down's syndrome in Dutch mainstream education. *European Journal of Special Needs Education 14*(3), 212-220.

Scheepstra, A. J. M., Pijl, S. J., and Nakken, H. (1996). "Knocking on the school door": Pupils in the Netherlands with Down's syndrome enter regular education. *British Journal of Special Education, 23*(3), 134-139.

*Schramm, B. J. (1974). *Case studies of two Down's syndrome children functioning in a Montessori environment. Research Project* (master's thesis). University of Dayton. Retrieved from ERIC database. (ED111120).

Sharpe, M. N., York, J. L., and Knight, J. (1994). Effects of inclusion on the academic performance of classmates without disabilities: A preliminary study. *Remedial and Special Education, 15*(5), 281-287.

*Shevlin, M., Walsh, P., Kenny, M., McNeela, E., and Molloy, R. (2003). Experiencing mainstream education: Exploring the perspectives of parents of young people who have Down syndrome. *REACH Journal of Special Needs Education in Ireland, 17*(1), 3-11.

Silva, B. K. L. N. (2009). *Inclusão escolar de uma criança com síndrome de Down* (in Portuguese) [*The school integration of a child with Down syndrome*]. Paper presented at IX Congresso Nacional de Educação – Educere III Encontro Sul Brasileiro de Psicopedagogia, 26-29 october 2009.

*Sinson, J. C., and Wetherick, N. E. (1981). The behaviour of children with Down syndrome in normal playgroups. *Journal of Mental Deficiency Research, 25*(2), 113-120.

Sinson, J. C., and Wetherick, N. E. (1986). Integrating young children with Down's syndrome. Gaze, play and vocalisation in the initial encounter. *The British Journal of Mental Subnormality, 32*(63), 93-101.

Sioutis, S. (2007). *Συνδρομο Down και οικογενεια στην Ελλαδα. Η παροχή κοινωνικών και εκπαιδευτικών υπηρεσιών στα άτομα με ειδικές αναγκες και τις οικογένειές τους. Μια διεπιστημονική προσέγγιση* (in Greek) [*Down syndrome and their families in Greece. Providing social and educational services to people with disabilities and their families. A multidisciplinary approach*] (PhD thesis). Panteion University, Athena, Greece.

Sirlopú, D., González, R., Bohner, G., Siebler, F., Ordóñez, G., Millar, A., Torres, D., and deTezanos-Pinto, P. (2008). Promoting positive attitudes toward people with Down syndrome: The benefit of school inclusion programs. *Journal of Applied Social Psychology, 38*(11), 2710-2736.

*Sloper, P., Cunningham, C., Turner, S., and Knussen, C. (1990). Factors related to the academic attainments of children with Down's syndrome. *The British Journal of Educational Psychology, 1990*(3), 284-298.

*Sloper, P., Turner, S., Knussen, C., and Cunningham, C. (1990). Social life of school children with Down's syndrome. *Child: Care, Health and Development, 16*, 235-251.

Solórzano Arriaga, M. Z. (2005). Integración social de alumnas con síndrome de Down en la educación regular* (in Spanish) [*Social integration of students with Down syndrome in regular education*] (PhD thesis). Universidad de San Carlos de Guatemala.

Staub, D. (2005). *Inclusion and the other kids. Here's what research shows so far about inclusion's effect on nondisabled students*. Arizona: Arizona State University, National Institute for Urban School Improvement (NIUSI).

Staub, D., and Peck, C. A. (1994/1995). What are the outcomes for nondisabled students? *Educational Leadership, 52*(4), 36-41.

Tanti Burlò, E. (2010). Inclusive education: a qualitative leap. *Life Span and Disability, 13*(2), 203-221.

Turner, S., and Alborz, A. (2003). Academic attainments of children with Down's syndrome: A longitudinal study. *The British Journal of Educational Psychology, 73*(4), 563-584.

*Turner, S., Alborz, A., and Gayle, V. (2008). Predictors of academic attainments of young people with Down's syndrome. *Journal of Intellectual Disability Research, 52*(5), 380-392.

Velez, L., Álvarez, I., and Ochoa, M. (1999). *Incidencia de los programas de estimulación en niños con síndrome de Down que se encuentran integrados al preescolar* (in Spanish) [*Effect of stimulation programs for children with Down syndrome that are integrated into preschool*]. Universidad de Antioquia. Colombia. Servicio de REDUC Número Rae 03.386-10.

*Vervat, B. (2008). *Competentiebeleving en Sociale Redzaamheid van kinderen met Downsyndroom. Regulier onderwijs versus speciaal onderwijs* (in Dutch) [*Self-perceived competence, social and self-help skills of children with Down syndrome. Regular versus special education*] (master's thesis). University Utrecht.

Vianello, R., and Lanfranchi, S. (2009). Genetic syndromes causing mental retardation: Deficit surplus in school performance and social adaptability compared to cognitive capacity. *Life Span and Disability, 12*(1), 41-52.

Vianello, R., and Moalli, E. (2001). Integrazione a scuola: le opinioni degli insegnanti, dei genitori e dei compagni di classe (in Italian) [Integration at school: the view of teachers, parents and classmates]. *Giornale Italiano delle Disabilità, 1*(2), 29-43.

Waldron, N., and Cole, S. (2000). *The Indiana inclusion study: Year one report.* Bloomington, Indiana: Indiana University, Indiana Institute on Disability and Community.

Wang, M. C., and Baker, E. T. (1985/86). Mainstreaming programs: design features and effects. *The Journal of Special Education, 19*, 503-521.

Wang, X., Bernas, R., and Eberhard, P. (2001). Effects of teachers' verbal and non-verbal scaffolding on everyday classroom performances of students with Down syndrome. *International Journal of Early Years Education, 9*(1), 71-80.

Wendelborg, C., and Tøssebro, J. (2010). Marginalisation processes in inclusive education in Norway: a longitudinal study of classroom participation. *Disability and Society, 25*(6), 701–714.

West, R. F., and Cummins, P. (1990). *Personal and professional change associated with the integration of children with Down syndrome into the public schools.* Paper presented at the Annual Meeting of the Mid-South Educational Research Conference (New Orleans, LA, November 14, 1990). Retrieved from ERIC database. (ED329093).

*Wilson, C. (2007). *Self-esteem in children with Down syndrome in the mainstream classroom.* Institute of Work, Health and Organisations, University of Nottingham. http://www.inclusive-solutions.com (retrieved 30-11-2011).

*Wolpert, G. (1996). *The educational challenges inclusion study.* New York: National Down Syndrome Society.

Wolpert, G. (2001a). What general educators have to say about successfully including students with Down syndrome in their classes. *Journal of Research in Childhood Education, 16*(1), 28-38.

Wolpert, G. (2001b). *Successful daily practices of inclusion teachers of children with Down syndrome.* Retrieved from ERIC database. (ED453160).

Wybranski, N. (1997). *The importance of multiple treatments in combination for the mild to moderately involved.* Paper presented at the Annual Conference of the Eastern Educational Research Association (February 19-22, 1997).

*Yadarola, M. E. (1996). *Síndrome de Down Estudio SocioPedagógico* (in Spanish) [*Down syndrome: a sociopedagogical study*] (PhD thesis). Universidad Nacional de Córdoba.

*Yadarola, M. E. (1998). Investigación comparativa sobre niños con síndrome de Down escolarizados en escuelas comunes y en escuelas especiales (in Spanish) [Comparative research on Down syndrome children enrolled in regular schools and special schools]. *Boletín de la Asociación Síndrome de Down de la República Argentina, 10*(31), 1-12.

York, J., Vandercook, T., Macdonald, C., Heise-Neff, C., and Caughey, E. (1992). Feedback from teachers and classmates about integrating middle school learners with severe disabilities in regular classes. *Exceptional Children, 58*(3), 244-259.

* Indicates all comparative and non-comparative studies included and reflected in Tables 1 and 2.

In: Advances in Genetics Research. Volume 12
Editor: Kevin V. Urbano

ISBN: 978-1-63321-351-7

Chapter 3

THE BIOLOGY OF TELOMERES IN MULTIPOTENTIAL STROMAL CELLS

Dimitrios Kouroupis, Sarah M. Churchman and Elena A. Jones

Leeds Musculoskeletal Biomedical Research Unit, Leeds Institute of Molecular Medicine, University of Leeds, Leeds, UK

ABSTRACT

Multipotential stromal cells, also known as mesenchymal stem cells (MSCs), reside within the bone marrow (BM) and other skeletal and connective tissues. They serve as a reserve of progenitors for continuous cell replacement during normal tissue homeostasis. Furthermore, MSCs are responsible for production of replacement cells lost as a result of acute or chronic injury to bone, cartilage, tendon, meniscus and other connective tissues and are involved in physiological bone remodeling. Loss of MSC function is apparent in the older individual, potentially contributing to the development of osteoporosis and osteoarthritis; dyskeratosis and Werner's syndromes are also linked to premature aging of skeletal and connective tissues. Telomere attrition has been proposed as a possible mechanism for the loss of MSC function, and may represent a therapeutic target in these diseases.

MSCs can be culture-expanded with relative ease and are already used therapeutically to repair cartilage and bone defects following injury. Also, because of their well-described immunoregulatory capacity, MSC therapy is used to treat patients with graft-versus-host disease. It is therefore very important to monitor telomere dynamics during MSC manufacture, for the purpose of controlling their therapeutic potency and potential for spontaneous transformation.

We will outline the current understanding of MSC ageing *in vitro* and observed telomere loss during culture expansion, primarily focusing on MSCs derived from adult tissues (BM, adipose and dental pulp), but also discussing perinatal tissues such as umbilical cord. Various telomere length measurement methods will be described and compared, in terms of their utility and practicality. Safety concerns relating to genetic manipulation of MSCs to extend their proliferation span, and the subsequent quantity of manufactured MSCs will be discussed.

Finally, we evaluate *in vivo* MSC ageing compared to that of hematopoietic stem cells highlighting both similarities and differences, which in our view stems from the current lack of robust methods for native MSC isolation directly from skeletal tissues. This combined knowledge should undoubtedly advance our understanding of MSC function during ageing and is likely to contribute to the development of novel therapies for age-related skeletal diseases.

INTRODUCTION

Multipotential Stromal Cells

Multipotential stromal cells (MSCs), also known as mesenchymal stem cells, were originally discovered in the bone marrow (BM) of adult animals and humans, and shown to differentiate into several skeletal and connective tissues including bone, cartilage, fibrous tissue and fat in *in vivo* animal models (Friedenstein et al. 1982; Friedenstein et al. 1987; Friedenstein et al. 1992). Subsequently progenitor cells with similar capabilities were found in other tissues, including subcutaneous adipose tissue, synovial tissue, periosteum, bone, skin and joint fat (Covas et al. 2008). Furthermore, similar cells were isolated from a number of perinatal tissues (umbilical cord, placenta) (Weiss et al. 2006; Karahuseyinoglu et al. 2007; Pasquinelli et al. 2007; Ishige et al. 2009; Fan et al. 2011) and embryonic tissues (fetal bone, fetal blood, fetal bone marrow) (Pasquinelli et al. 2007; Guillot et al. 2008) as well as from amniotic fluid (Bajada et al. 2008; Sessarego et al. 2008). It is now well recognized that MSCs have considerable tissue-specificity (Sakaguchi et al. 2005; De Bari et al. 2008; Jones et al. 2008), however they do possess a number of common characteristics which were used to define MSCs by the International Society of Cellular Therapy (ISCT) (Dominici et al. 2006). According to the ISCT criteria MSCs must be adherent to plastic *in vitro*, express several positive markers (CD13, CD44, CD73, CD90 and CD105) and lack expression of haematopoietic markers (CD14, CD11b, CD79, CD34, CD45 and HLA-DR) (Dominici et al. 2006). Additionally, MSCs must be able to differentiate at least to osteoblasts, adipocytes and chondroblasts under standard conditions *in vitro* (Dominici et al. 2006). Due to their residence in connective tissues composed predominantly of fibroblasts (Sudo et al. 2007; Covas et al. 2008), MSCs and fibroblasts share common characteristics including the expression of several surface and molecular markers (Jones et al. 2004; Ishii et al. 2005), however common skin fibroblast cultures are considerably inferior to MSCs in terms of their multipotentility (Alt et al. 2011).

Telomeres in Haematopoietic Stem Cells (HSCs) – Gold Standard for MSC Research in this Area

Telomere biology in haematopoietic progenitor/stem cells (HSCs) and their progeny are much better understood compared to MSCs. From as early as the late 1960s, HSCs have been routinely transplanted into patients suffering from leukaemia and other BM failure syndromes and there was a strong clinical need to establish whether donor HSCs age and die, or conversely, function normally in transplant recipients (Ogden and Micklem 1976). HSCs give rise to numerous blood cell lineages including red cells, platelets and different types of leukocytes (lymphocytes and granulocytes). HSCs are normally isolated from the BM based on the expression of the CD34 marker (Civin et al. 1996) and leukocytes are isolated easily from peripheral blood using differential density centrifugation. A number of studies compared HSCs and their progeny; leukocytes, to investigate whether the telomere length of these cells could be used as a marker of HSC 'immaturity' and/or HSC potential for life-long reconstitution in transplant patients. In aged donors, HSCs possess reduced repopulation potential which is currently explained by general exhaustion of the HSC pool due to continuous HSC turnover in an individual's lifetime (Kamminga et al. 2006; Glauche et al. 2011).

Telomere length is heritable and is further modified in an embryo by transient activation of a number of telomere lengthening mechanisms. In newborn umbilical cord blood $CD34^+$ HSCs, telomere lengths were found to range between 8 and 10.5kb (Kimura et al. 2010). This is lower than the average telomere length range in germline cells (sperm, range 10-20kb) (Allsopp et al. 1992) and higher compared to somatic cells (fibroblasts, range 6-10kb) (Allsopp et al. 1992). These findings indicate that although relatively long telomeres are a characteristic feature of adult stem cells including HSCs (Flores et al. 2008; Kimura et al. 2010), it is possible for telomeres to be long in more mature cell types, depending on the individual. This also highlights another important point that has further relevance to MSCs: that the magnitude of inter-individual variation in telomere lengths within the same cell type can be considerable and in fact can exceed the variation between different cell types within the same person (Friedrich et al. 2000; Takubo et al. 2002; Gadalla et al. 2010).

Since leukocyte telomere length (LTL) closely mirrors that of HSCs (Kimura et al. 2010), telomere length heritability has most commonly been investigated in leukocytes. Approximately 80% inheritance has been reported in both monozygotic and dyzygotic twins (Slagboom et al. 1994; Jeanclos et al. 2000). Although the mode of LTL inheritance has been suggested to be X-linked (Nawrot et al. 2004), a recent study also shows a paternal mode of heritability (Nordfjall et al. 2009). Telomere loss in lymphocytes and granulocytes occurs rapidly within the first two years of life (Rufer et al. 1999) and afterwards it continues to decline for more than eight decades at a much lower rate (Rufer et al. 1999; Oeseburg et al. 2010). In the thymus, T lymphocyte telomeres are believed to be longer than in granulocytes (Halaschek-Wiener et al. 2008; Kimura et al. 2010). Interestingly, in a group of healthy "oldest old" (over 85 years old), reduced variation in LTL was found, suggesting that average/optimal telomeres may be protective, whereas short or long telomeres could both represent risk factors for tumour development (Halaschek-Wiener et al. 2008). Various lifestyle factors including exercise, smoking and diet can be also associated with telomere dynamics in leukocytes (Cassidy et al. 2010), with longer telomeres being linked with a non-stressful lifestyle (Epel et al. 2004), increased physical activity (Cherkas et al. 2008) and a healthier diet (Cong et al. 2002).

Aging and Telomere Shortening Following Standard in Vitro Expansion of BM MSCs

The study of telomere dynamics in MSCs and their progeny represents a more considerable challenge compared to HSCs. This is because the native phenotype of MSCs in different human tissues has not been fully elucidated or completely agreed upon (Jones and McGonagle 2008; Bianco et al. 2010; Jones and Yang 2011). Some consensus currently exists for the BM MSC phenotype; they are known to express low levels of CD45 and high levels of CD271, CD146, CD105 and CD73 (Buhring et al. 2009; Jones and Yang 2011; Tormin et al. 2011). Other selective markers for BM MSCs have been proposed in the past (Buhring et al. 2007; Gang et al. 2007; Martinez et al. 2007), but their utility in comparison to CD271, CD146, CD105 or CD73 remains to be elucidated. In human adipose tissue, MSCs have been reported as $CD271^+CD34^+$ (Quirici et al. 2010) however this fraction may not solely contain MSCs. Also, in human umbilical cord tissue, MSCs have been described as $CD146^+$ but this fraction may similarly be not exclusive to MSCs (Baksh et al. 2007; Schugar et al. 2009) as CD146 is additionally expressed on the endothelial cells and some leukocytes (Elshal et al. 2005). In general, CD146 has been shown to be expressed on MSCs obtained from diverse human tissues (Crisan et al. 2008).

In the absence of a unanimously-agreed set of markers to define native MSCs in diverse human tissues, their telomere biology *in vivo* has not been investigated in the same depth as that of HSCs and the overwhelming majority of studies have been done on MSC cultures expanded *in vitro*.

Telomere dynamics in cultured MSCs is most comprehensively studied in BM MSCs; these cells are normally expanded as high-density cultures in media containing foetal calf serum and no or very little supplementation with specific growth factors (Pittenger et al. 1999; Jones et al. 2002; Tarte et al. 2010). These basic conditions support MSC self-renewal to only a limited degree, resulting in late-passage cultures being depleted of true MSCs and subsequently enriched for pre-senescent, less proliferative cells (Wagner et al. 2008). It is generally accepted that MSC cultures commit to an osteogenic lineage and acquire strong signs of senescence after 20-40 population doublings (PDs) in standard culture (Bruder et al. 1997; Muraglia et al. 2000; Wagner et al. 2008). In relation to telomere dynamics, this "*in vitro* MSC ageing" was first documented by Banfi et al., who showed that standard BM MSC cultures undergo telomere attrition at the rate of 50-100bp per PD (Banfi et al. 2002). In agreement, later studies similarly reported progressive telomere erosion in human BM MSCs, with no telomere maintenance mechanisms or cellular transformation observed in late cultures (Baxter et al. 2004; Bonab et al. 2006; Bernardo et al. 2007; Jones et al. 2010). The telomere loss in MSCs *in vitro* appears to be linked with their general loss of multipotentiality (Banfi et al. 2000; Stolzing et al. 2008). It should be noted that MSC aging *in vitro* depends on the species from which the cells are derived. Human BM MSCs show replicative senescence between 30 and 50 PDs (Stenderup et al. 2003) whereas murine BM MSCs can be culture expanded for more than 100 PDs (Meirelles and Nardi 2003).

Baxter et al., have reported that human BM MSC cultures stop dividing when their telomere length reaches approximately 10kb (Baxter et al. 2004). The implication of this work is that, theoretically, the number of PDs that any human BM MSC culture can reach prior to senescence can be deduced from the mean telomere length in these cells at an early passage (Baxter et al. 2004). Similar rates of telomere attrition were observed in Mareschi et al., however, the telomere lengths in senescent MSCs were found to be as low as 8kb (Mareschi et al. 2006). Of note, upon *in vitro* cultivation of HSCs ($CD34^+$ cells), the observed telomere attrition rates were found to be similar (Vaziri et al. 1994).

The Relationship between Telomere Length in BM MSCs and Donor Age

Several studies have shown that BM MSCs from younger donors grow faster and for longer than those of aged donors (Mareschi et al. 2006; Stolzing et al. 2008; Jones et al. 2010). One hypothesis postulates that this could be a result of longer telomeres in culture-initiating native MSCs from younger individuals (Baxter et al. 2004). By extrapolating their *in vitro* findings to the *in vivo* situation, Baxter et al. calculated that an average telomere loss in BM MSC *in vivo* could be as high as 17bp/year of life (Baxter et al. 2004). Using a small cohort of donors and the $CD45^{-/low}CD271^{+}$ phenotype to define native BM MSCs, a trend for shorter telomeres in the BM of older donors has been observed in one study (Jones et al. 2010); these findings are awaiting further independent confirmation. In the parallel HSC field, there remains a lack of clarity as to whether native HSCs and their progeny 'age' *in vivo*. Earlier studies reported age-dependent decline in telomere length in purified HSCs (Vaziri et al. 1994; Piacibello et al. 2005), however no significant decline in telomere length in $CD34^{+}$ HSCs was documented in a more recent study (Wagner et al. 2009). This can be explained by the use of different techniques to measure the lengths of telomeres. The rate of LTL shortening in adults has been described as ~30bp/year although this was variable between the individuals (Sidorov et al. 2009).

Besides an individual's age, telomere attrition rates in MSCs *in vivo* are likely to be determined by other factors, including the demands for growth, bone and marrow tissue turnover at different stages of life, and environmental factors as described earlier for HSCs. MSCs give rise to osteoblasts and osteocytes, cells that form new bone (Parfitt 1984). Bone turnover takes place throughout life and each bone remodelling cycle is usually completed every four months (Parfitt 1984). Whereas HSCs are responsible for generating 200 billion erythrocytes and 70 billion neutrophilic leukocytes daily (Gunsilius et al. 2001), skeletal tissue turnover, with complete replacement of bone-forming cells, is estimated to be once a decade (Parfitt 2002). Higher bone turnover rate can actually lead to the weakening of bone, disruption of trabecular bone structure and increased risk of fracture (Dempster 2002). Therefore the rate of ageing and potential telomere loss in native MSCs is likely to be much slower compared to that of HSCs. As expected, fetal MSCs have longer telomeres compared to adult MSCs from the same tissue (Guillot et al., 2007), but it should be emphasized that these studies have been performed on MSCs expanded in culture, which in itself leads to telomere attrition, as described above.

Improved Telomere Maintenance in BM MSCs Grown in Modified Culture Expansion Conditions

Several modifications in basic MSC expansion conditions have been shown to prolong the lifespan of BM MSC cultures (Sekiya et al. 2002). These modifications include the use of batch-tested foetal calf serum, basal medium, stable glutamine and glucose concentrations, cell plating and splitting densities and plastic surface quality. Lower cell densities, which allow cells to proliferate, favour osteogenic differentiation whereas higher densities that cause cell condensation favour adipogenic commitment (Sekiya et al. 2002; McBeath et al. 2004). Chondrogenesis also increases with higher cell density leading to greater cell condensation (Mackay et al. 1998; Gregory et al. 2005).

Early and more recent studies showed that the addition of several growth factors, notably basic fibroblast growth factor (bFGF), can lead to lesser telomere attrition and prolong the longevity of human BM MSC cultures to 70-100 PDs (Bianchi et al. 2003; Sotiropoulou et al. 2006; Yanada et al. 2006). Culturing BM MSCs in media supplemented with low-concentration FGF is clinically safe and has been used for the manufacture of MSCs in clinical trials (Tarte et al. 2010). Earlier studies have employed low-serum media supplemented with epidermal growth factor and platelet-derived growth factor BB, fibronectin-coated plastic and low seeding density to derive MSC-like cultures with improved longevity and broader differentiation potentials (termed marrow adult progenitor cells, MAPCs) (Reyes et al. 2001). The average telomere length of late-passage MAPCs (35 PDs) was very long (between 11-15kb), approximately 3kb longer than in leukocytes from the same donors (Reyes et al. 2001). The telomere length of MAPCs from one donor, tested after 15, 35, and 45 cell PDs remained unchanged, suggesting that the chosen culture conditions supported telomere maintenance in MAPCs by as yet unknown mechanisms (Reyes et al. 2001).

Another group of scientists have derived pluripotent cells by growing BM adherent and non-adherent fractions on fibronectin and at low oxygen tension levels (termed marrow-isolated adult multilineage inducible (MIAMI) cells (D'Ippolito et al. 2004). Although telomere dynamics in these culture-derived cells have not been evaluated, the authors demonstrated that the frequency of MIAMI cells, among all marrow nucleated cells, decreased from 0.01% at age 3 years to 0.0018% at 45 years, and their numbers remained unchanged after the age of 45 (D'Ippolito et al. 2004). It was suggested that culturing MSCs in low oxygen conditions closely mimicked their residence near bone surfaces. In agreement, Tormin et al., 2011 have recently shown that $CD271^{+}$ MSCs were located in both perivascular and bone-lining regions (Tormin et al. 2011). These data further highlighted the need to study telomere dynamics in freshly-isolated rather than culture expanded MSCs as it is clear that culture conditions profoundly affect longevity and telomere maintenance in these cells.

Finally, forced telomerase (TERT) expression in MSCs has been unsurprisingly shown to extend the lifespan and enhance differentiation potential of MSCs (Simonsen et al. 2002). MSC-TERT has subsequently been used as a valuable tool in MSC proteomics (Foster et al. 2005) and for the study of MSC engraftment *in vivo* (Bentzon et al. 2005). The pattern of MSC distribution in the organs suggested limited ability of MSCs to be transplanted systemically, with the majority of infused cells being arrested in microvasculature of lungs (Bentzon et al. 2005), consistent with similar reports (Prockop 2009). The therapeutic utility of these cells however remains very limited, as long-term cultures of MSC-TERT have shown to evolve spontaneous genetic changes leading to tumorigenicity in mice (Burns et al. 2005).

Telomere Dynamics in MSCs from Synovium and Periosteum

BM-derived MSCs represent an excellent source of reparative cells in bone repair (Quarto et al. 2001) due to their natural role in bone remodeling *in vivo* and also, as already mentioned, their osteogenic commitment capability during standard *in vitro* expansion (Muraglia et al. 2000). BM MSCs are believed to percolate from the marrow cavity and thus contribute to the repair of full-thickness cartilage defects during a common orthopedic procedure called microfracture (Endres et al. 2010). Partial-thickness cartilage defects do not breach subchondral bone and thus the regeneration of these defects is more challenging, with MSCs from synovium (De Bari et al. 2001; Lee et al. 2010; Mendelson et al. 2011), periosteum (Fukumoto et al. 2003; De Bari et al. 2006) or synovial fluid (Jones et al. 2004; Jones et al. 2008) being considered as most plausible alternatives. MSCs derived from synovial membrane (SM MSCs) and periosteum (P MSCs) were reported to be present at 10 to 100-fold higher frequency compared to BM MSCs (Sakaguchi et al. 2005; De Bari et al. 2006). Early-passage P MSCs exhibited reasonably long telomeres (up to 9kb), with a trend for shorter telomeres in MSCs from the oldest donors (De Bari et al. 2006). In a subsequent study, telomere lengths in SM MSCs and P MSCs from the same 4 donors were compared and found to be similar (11.5kb and 11.45kb, respectively) but not age-dependant (De Bari et al. 2008). When SM MSCs and P MSCs were grown as clonal cultures (derived from single ancestral cells), mean telomere lengths were found to be 7.08kb and 7.53kb, respectively (De Bari et al. 2008). These results highlight the fact that culture conditions and MSC replicative history *in vitro* can affect the length of telomeres in MSCs, regardless of their tissue of origin. It is likely that clonal SM MSCs and P MSCs have undergone more divisions compared to standard SM and P MSC cultures prior to the telomere length analysis resulting in the shorter telomere lengths observed.

Friedrich *et al.*, investigated telomere lengths in the matched tissues (leukocytes, skin and synovial tissue) of nine elderly donors (Friedrich et al. 2000). They observed significantly shorter telomeres (6.5kb) in leukocytes (fast replicating cells) compared to skin (7.8kb) and synovial tissue cells (7.9kb). The authors speculated that skin and synovial tissue are likely to contain slower replicating cells compared to leukocytes, thus explaining longer telomeres in these tissues (Friedrich et al. 2000). However, since significant negative relationships were observed between telomere lengths in all three tissues examined and the age of the donors, the authors proposed that there is "a common, probably genetically controlled mechanism" that likely controls the regulation of telomeres in various cell types.

The *in vivo* phenotype of synovial MSCs remains unknown. Whilst the majority of studies resort to using surface markers to purify native MSCs, an entirely novel approach was undertaken in a recent study by Kurth *et al.,* 2011. This was based on a concept that tissue-resident stem cells are quiescent (or slow-cycling), however they can be prompted to proliferate and differentiate in response to injury to that tissue. In this study, a double nucleoside-analogue cell-labelling scheme was used in a mouse model of joint surface injury (Kurth et al. 2011). Slow-cycling cells were found to be non-haematopoietic, non-endothelial stromal cells and stained positive for MSC markers. After injury, these cells underwent proliferation and differentiation, consistent with their nature as synovial MSCs (Kurth et al. 2011). It would be very interesting to determine telomere lengths in these cells.

Adipose-Derived MSCs

Autologous adipose tissue (AT) derived MSCs have been used for many therapeutic applications including spinal fusion, soft tissue and bone repair (Helder et al. 2007; Seong et al. 2010). AT MSCs are quite abundant in lipoaspirates and fatty solid tissues (1-3%) (Zuk et al. 2002; Mizuno 2009). AT MSCs are multipotential towards bone, fat and cartilage *in vitro* (Zuk et al. 2002) and support osteogenesis, vasculogenesis and neuronal repair *in vivo* (Cousin et al. 2003; Kang et al. 2003; Cowan et al. 2004). Upon expansion, AT MSC cultures enter growth arrest after 60 to 80 PDs (Vidal et al. 2011) and as with osteoblasts, pre-senescent cells gradually lose their differentiation capacities (Noer et al. 2007). Kang et al., have demonstrated very long telomeres (23 kb) in early AT MSC cultures from non-human primates (Kang et al. 2004). In Izadpanah et al., rhesus monkey BM MSCs and AT MSCs indeed had longer telomeres than human BM and AT MSCs and no significant reduction in telomere length was observed during culture expansion of AT MSCs from both sources (Izadpanah et al. 2006).

Mouse AT MSCs has been found to display an age-dependent decrease in proliferation and telomere length when aged and young animals were compared (Efimenko et al. 2011). More recently, telomere length, telomerase activity and osteogenic differentiation were found to be similar in healthy and osteoporotic murine AT MSCs (Mirsaidi et al. 2011). AT MSCs overexpressing TERT retained their multilineage potency towards adipogenic, chondrogenic and neurogenic lineages, late in culture (Wolbank et al. 2009). Whether or not culture expanded human AT MSCs age at the same rate as BM MSCs, needs to be investigated further using adipose samples from different anatomical areas. It is well recognized that the biological characteristics of AT MSCs differ depending on the body region from which the adipose tissue was collected (Levi et al. 2010).

Umbilical Cord-Derived MSCs

Umbilical cord (UC) tissue originates from the extraembryonic mesoderm at day 13 of embryonic development (Karahuseyinoglu et al. 2007). Full-term UC was found to be rich in MSCs (up to 5% of total cells) (Majore et al. 2011) and easily accessible for MSC harvesting. UC MSCs have potential for cell-based therapy, primarily based on their stimulatory and modulatory effects on host regenerative cells post-transplantation (Weiss et al. 2006; Yang et al. 2008). UC MSCs meet the basic ISCT criteria (Dominici et al. 2006) for MSCs (Baksh et al. 2007). Besides the standard tri-lineage differentiation capacity, UC MSCs have been shown to differentiate towards cardiomyocytes (Wang et al. 2004), dopaminergic neurons (Weiss et al. 2006), islet-like cells (Wu et al. 2009) and hepatocytes (Campard et al. 2008).

UC MSCs possess high expansion potential (above 60 PDs) (Can and Karahuseyinoglu 2007; Moretti et al. 2010) and high chromosomal stability (Rebuzzini et al. 2008). The high proliferative capacity of UC MSCs is believed to be linked with high expression of embryonic stem cell (ESC) markers including Oct-4 and Nanog (Carlin et al. 2006; Zhang et al. 2009). UC MSCs have been shown to have long telomeres (>9kb), both early (24 PDs) and late (60 PDs) in culture (La Rocca et al. 2009). Telomere stability in UC MSCs can be a result of their higher telomerase activity compared to somatic cells in the body (Jo et al. 2008). Interestingly, comparison of UC with cancer cell lines revealed higher telomere lengths (7.2-9.3 kb versus 3.9-6 kb, respectively) but lower telomerase activity in MSC cultures suggesting that alternative telomere lengthening mechanisms may be activated in MSCs (Jeon et al. 2011). In one study, telomere length in UC MSCs was shown to be as high as 20kb (Fong et al. 2010).

UC blood (UCB) has large quantities of HSCs (Cohen and Nagler 2004) and relatively small quantities of MSCs (Kögler et al. 2004; Moretti et al. 2010). UCB MSCs are tripotential *in vitro* (Erices et al. 2000; Goodwin et al. 2001), but compared to UC MSCs exhibit earlier growth arrest (after 30 PDs) (Terai et al. 2005). Interestingly, Pacini et al., identified two distinct cell populations in UCB MSCs: $SSC^{low}SSEA\text{-}4^{neg}CD105^{bright}MSCA\text{-}1^{+}CD90^{bright}$ and $SSC^{high}SSEA\text{-}4^{+}CD105^{dim}MSCA\text{-}1^{neg}CD90^{neg}$, showing 5.88kb and 9kb length telomeres, respectively (Pacini et al. 2010). The limited lifespan of UCB MSCs can be prolonged for >120 PDs using a telomerase transfection technique (Terai et al. 2005). Finally, Oktar et al. have recently reported that nucleostemin nucleolar protein is involved in telomere maintenance in UCB MSCs showing gradually decreased expression upon their aging *in vitro* (Oktar et al. 2011).

Dental Pulp-Derived MSCs

Dental mesenchyme originates from ectomesenchyme; therefore dental MSCs display many characteristics of neural stem cells (Huang et al. 2009). Dental pulp MSCs (DP MSCs) were first isolated from wisdom teeth and reported to form dentin/pulp-like structures both *in vitro* and *in vivo* (Gronthos et al. 2002). Subsequently, four more types of dental-MSC-like populations were isolated and characterized: exfoliated deciduous teeth stem cells (Miura et al. 2003), periodontal ligament stem cells (Seo et al. 2004), dental follicle precursor cells (Morsczeck et al. 2005) and apical papilla stem cells (Sonoyama et al. 2006). The precise relationship between these dental stem cell types remains unclear.

DP MSCs are the most characterized dental-derived MSC type. They are plastic-adherent, express MSC-specific markers and exhibit high proliferative capacity (Gronthos et al. 2002; Miura et al. 2003; Huang et al. 2009). DP MSCs can differentiate into dentin-producer odontoblasts (Gronthos et al. 2002; Iohara et al. 2004; Zhang et al. 2006), osteoblasts (Laino et al. 2005; Zhang et al. 2006; d'Aquino et al. 2007), adipocytes (Gronthos et al. 2002), chondroblasts (Zhang et al. 2006), smooth muscle cells (Gandia et al. 2008), endothelial cells (d'Aquino et al. 2007) and neural cells (Gronthos et al. 2002; Arthur et al. 2008). DP SCs have higher telomerase levels compared to BM MSCs (Sonoyama et al. 2006) and can be expanded for more than 81 PDs (Mokry et al. 2010). Similar to other types of MSC, DP MSCs display progressive telomere shortening in long-term culture *in vitro* (Mehrazarin et al. 2011; Mokry et al. 2010), which appears to correlate with donor age, showing shorter telomere lengths in elders (Mokry et al. 2010).

Human Skeletal Diseases Associated with Excessive or Accelerated Telomere Shortening

Telomere shortening due to defective telomerase function or mutations in the DNA repair system has been described to be a feature of many skeletal diseases. Dyskeratosis congenita (DC) is a premature aging syndrome associated with mutations affecting the telomerase complex resulting in shorter telomeres and decreased telomerase stability (Nelson and Bertuch 2011; Mitchell et al. 1999). DC is a clinically heterogeneous inherited BM failure syndrome involving different pathologies such as developmental delay, cerebellar hypoplasia, avascular necrosis of the hips or shoulders, pulmonary fibrosis, bone marrow and haematopoietic defects, and susceptibility to malignancies (Walne et al. 2005; Yamaguchi 2007). The combination of pathologies related to both haematopoietic and skeletal systems in these patients suggests that both HSCs and MSCs may be affected. Similarly to DC, aplastic anemia is linked to telomerase complex mutations and characterized by immune-mediated BM failure (Marrone et al. 2004).

Werner syndrome (WS) is another premature aging syndrome associated with mutations in genes of the DNA repair system showing accelerated telomere loss in systemic organs (Ishikawa et al. 2011). Clinical manifestations of WS include atherosclerosis, diabetes mellitus, cataracts, menopause and susceptibility to malignancy (Goto 2000). Osteoporosis (OP) is common in both Werner and DC premature aging syndromes (Pignolo et al. 2008) suggesting that premature ageing in MSCs may be an important contributing factor in the pathogenesis of these diseases. Telomere length analysis in MSCs from these patients is awaiting future investigation.

Osteoarthritis (OA), rheumatoid arthritis (RA) and OP have been also associated with replicative senescence caused by telomere shortening. In RA, it is chronic inflammation and oxidative stress that most likely contribute to telomere shortening in leukocytes (Goronzy and Weyand 2003). Cultured BM MSCs from RA patients have also been shown to have accelerated telomere loss compared to age-matched controls (Kastrinaki et al. 2006), but the mechanism of this remains unknown. OA is a progressive degenerative disease that mostly affects cartilage, with senescent cells having been detected in osteoarthritic lesions located in chondrocyte clusters with high mitotic activity. In normal cartilage chondrocyte turnover is low, but during disease progression high replicative activity and oxidative stress cause rapid telomere shortening and chondrocyte death (Price et al. 2002). No differences in telomere length in iliac crest BM MSCs, an area not affected by disease, were found in OA patients compared to age-matched controls in one study (Jones et al. 2010). Similarly, LTL in OA patients showed no difference from age-matched controls (Tamayo et al. 2010). In contrast, LTL in OP was significantly shorter compared to controls (Valdes et al. 2007), however no such abnormality was observed in another study (Sanders et al. 2009). In all of these diseases, the investigation of telomere dynamics in native MSCs, in areas remote to and adjacent to damage sites needs to be explored, to fully elucidate the role of premature MSC ageing in these diseases.

Measurement of Telomeres in MSCs

Since it is believed that telomere length is a surrogate measure of the replicative history (Hills et al. 2009) and the potential longevity of cells, then it is important to establish reliable methods for measuring their length, particularly when considering the use of MSCs for therapeutic purposes. The methods available for this have, as with anything, evolved with the advent of new technology. Whilst the choice of technique is reasonably diverse, including single telomere length analysis (STELA) (Baird et al. 2003), quantitative-fluorescence in situ hybridisation (Q-FISH) (Lansdorp et al. 1996), flow-FISH (Schmid et al. 2002; Baerlocher et al. 2006), G-tail telomere hybridisation protection assay (Tahara et al. 2005), quantitative real-time polymerase chain reaction (qPCR)/monochrome multiplex (MM)QPCR (Cawthon 2002; Cawthon 2009), terminal restriction fragment (TRF) length analysis (Moyzis et al. 1988), and many other blot/probe based methods (Bryant et al. 1997; Carbonari et al. 2010; Kimura and Aviv 2011), the frequency, volume and number of test samples are often the determining factors.

The TRF method, first described by Moyzis et al., (Moyzis et al. 1988), has been the most frequently used technique in measuring the length of telomeres in cultured MSCs (Stenderup et al. 2003; Baxter et al. 2004; Bernardo et al. 2007; De Bari et al. 2008; Kim et al. 2009). This involves digestion of total genomic DNA with restriction enzymes, unable to cleave the telomere repeat region (TTAGGG). The resulting long telomeric and shorter remaining genomic fragments are resolved on agarose and then probed, using a telomere specific probe (CCCTAA repeats), following Southern blot. The resulting smear is used to calculate the mean telomere length of the sample, however this is open to interpretation as the result is not a discrete band or value. The choice of restriction enzyme can be that of the user and therefore results may not be directly comparable between laboratories.

Thus far in MSCs, it has been less common for mean telomere length to be elucidated by 3D-Telo-Q-FISH (Wagner et al. 2009) or qPCR (Guillot et al. 2007; Jones et al. 2010; Mokry et al. 2010), a more suitable choice for those working with limited cell numbers. Q-FISH, useful for small data sets, directly hybridises a $(CCCTAA)_3$ probe to the telomeres of metaphase cells and normalised to known values (Poon and Lansdorp 2001). It is useful for rare cells and can measure short (<500bp) repeat regions (Aubert et al. 2011). qPCR provides a normalised value of telomere repeats determined against a single copy gene (Cawthon 2002). STELA (Baird et al. 2003) would also provide a suitable alternative for research into rare cells (≥50 cells), and is arguably, along with Q-FISH (Lansdorp et al. 1996), the most accurate method in assessing shorter telomeres. STELA is performed using modified PCR targeting of a 'linker' hybridised to the XpYp chromosomes' telomeric region (Baird et al. 2003). The amplification products (of variable size) are then separated on agarose, Southern blotted and then probed using an XpYp subtelomeric probe (Hills et al. 2009). Telomere length is then calculated based on length of product compared to a DNA ladder and frequency/intensity of each probe/band on the blot. STELA has its own disadvantages, notably longer duration, it is also more technically demanding than qPCR, it only measures the telomere length of certain chromosomes and could not be considered high throughput. Both qPCR and STELA are limited in their ability to measure longer (>18) telomeres (Baird et al. 2003; Hills et al. 2009), a drawback when exploring MSCs and other adult stem cells which are expected to have longer telomeres. This has direct relevance to translational investigations where cell isolates with greater potential longevity and hence longer telomeres are desired.

Q-FISH, in particular high throughput-Q-FISH, has been developed for automated screening studies into cancer and age related studies (Canela et al. 2007), a scale not possible with STELA. Flow FISH (Rufer et al. 1998; Baerlocher et al. 2006), was developed as a modification of Q-FISH, measured using flow cytometry. This technique does require greater cell numbers than Q-FISH and the cells must be fresh, however, it can be automated. Flow FISH involves hybridization with labeled peptide nucleic acid probes, specific for telomere repeats, in combination with fluorescence measurements. Commonly, control 1301 cells of a known telomere length are used in the same tube to permit absolute telomere length measurements in kilobases (Schmid et al. 2002). Flow FISH has the unique advantage of providing multi-parameter information on the length of telomere repeats in thousands of, and simultaneously in, several cell populations (Baerlocher et al. 2006) different in their surface phenotypes. Very recently, Flow FISH was used to measure LTL in DC patients (Gadalla et al. 2011).

Since inter-laboratory variations can often account for discrepancies between studies, it is sometimes difficult to determine what is indeed a true result and what may be considered technical. When many of the earlier studies into the telomere length of blood/placenta/sperm (Allshire et al. 1989), tumor resections (Gertler et al. 2004), ESCs (Forsyth and McWhir 2008) and other tissues were performed the method protocol remained quite variable, not only in relation to the tissues investigated, but also the selection of restriction enzymes used. It was therefore debatable as to how comparable earlier findings from different laboratories were. On the other hand, the majority of MSC telomere length assessment has been carried out using a uniform method (the TRF/Southern blot method with HinfI/RsaI restriction enzymes) (Stenderup et al. 2003; Baxter et al. 2004; Bernardo et al. 2007; De Bari et al. 2008; Kim et al. 2009). It is therefore very likely that any differences observed between the MSC types studied or between different passages of same MSC cultures were truly reflective of the type of MSCs or culture conditions used. An example of this is well illustrated by Fong et al (Fong et al. 2010); they found that in Wharton's jelly-derived UC MSCs, despite p1 cultures having 20.2kb telomeres in different media, the telomere lengths became 16.5kb/14.6kb at p5/p10 when grown in complex medium, yet in simple medium the lengths were 15.4kb/12.5kb.

As mentioned earlier, the qPCR method of Cawthon (Cawthon 2002) was developed, normalizing the variable length telomere product to a fixed single copy number gene, as a rapid alternative to the RTF/Southern blot. Requiring only 20ng DNA per reaction as opposed to 0.5-10 μg for TRF (Aubert et al. 2011) and with its significantly shorter experimental time, it is on the surface a much more desirable technique, but this should be handled with care due not only to the sensitive nature of qPCR, but also with respect to the necessity of using the separate plate to derive the T/S ratio. The updated MMQPCR (Cawthon 2009) circumvents some of these concerns since both fixed copy and telomere gene products are amplified in the same reaction, generating the T/S ratio.

The correlations between telomere lengths derived by TRF and, qPCR and MMQPCR were found to be very good with correlation coefficients (r) of 0.823 and 0.919 respectively (Cawthon 2002; Cawthon 2009). The correlation between TRF and qPCR has also been verified independently ($r>0.9$) (Aviv et al. 2011). It is apparent that telomere lengths derived using qPCR are reported in the form of the T/S ratio (Guillot et al. 2007; Jones et al. 2010); this is a calculation point in Cawthon's original qPCR method and not the final calculated telomere length, which could in turn be derived. Whilst this may not enable immediate comparisons with studies utilizing the TRF method, both of the above-mentioned studies reported comparable T/S ratios for BM MSCs, confirming that both TRF and qPCR are compatible.

Since the source of sample can vary as can the scale and subject of the study being undertaken, there is not a universal 'best' method for measuring telomeres in MSCs (Aubert et al. 2011). However, with the reported native MSC frequencies and hence numbers of purified cells anticipated to be low, combined with the expectation that longer telomeres would be identified, it appears that a method capable of working at low cell numbers/DNA concentrations over a greater range of length such as qPCR/MMQPCR would be preferable. Conversely, if the study material is not scarce (such as cultured MSCs), TRF potentially represents a method of choice as it is robust and gives better coefficients of variation (1.75%) compared to the 6.45 of qPCR (Aviv et al. 2011).

CONCLUSION

As outlined in the above sections, the study of telomere biology in MSCs is in its infancy. The definition of MSCs remains based on the *in vitro* phenotype and differentiation capability of culture-expanded cells (Dominici et al. 2006). The lack of robust surface markets to purify native MSCs from diverse connective tissue hinders the progress in this area. The methods for isolation and culturing MSCs differ from source to source; the extraction of MSCs from solid tissues requires an enzymatic digestion step which is not needed when homogenous tissues like BM aspirates or UCB are used. However, the common theme that unifies all the tissues described in this chapter, is that despite obvious differences in MSC isolation, their telomeres shorten, regardless of the tissue type, when MSCs are cultured in basic standard conditions. This resembles the behaviour of common somatic cells (Fehrer and Lepperdinger 2005; Zou et al. 2009), including skins fibroblasts (Allsopp et al. 1992; Allsopp and Harley 1995) and does not significantly differ from other adult stem cells such as $CD34^{+}$ HSCs (Vaziri et al. 1994). In both cultured MSCs and HSCs telomere attrition is associated with the loss of proliferative capacity (Vaziri et al. 1994; Baxter et al. 2004). In MSCs, their differentiation capabilities also seem to be affected (Stolzing et al. 2006).

Culture-expanded MSCs are currently used for regenerating bone following fracture or bone tumour resection (Jones and Yang 2011) and are predicted to become a novel therapeutic option for cartilage repair in osteoarthritis (Barry and Murphy 2004). Based on their immunoregulatory capacity (Le Blanc et al. 2003), cultured MSC preparations are currently used as cell therapy to control graft-versus-host disease (Le Blanc et al. 2004). The current therapeutic exploitation of cultured MSCs does not take into consideration their proliferative history. Based on the literature described in the above sections, it is clear that all types of MSCs are artificially "aged" in standard culture conditions and are likely to contain large proportions of pre-senescent cells. This can significantly diminish repopulation and regeneration potentials of manufactured MSC batches.

It appears natural that the telomere length measurement in manufactured MSCs may be a key method of controlling the "quality" of cells. A critical telomere length in MSCs may be established beyond which MSC cultures would be deemed too "old" and rejected for transplantation. Although this sounds appealing, two major obstacles exist, in our opinion, that currently preclude the use of telomere length measurement as a biomarker of cell ageing in MSCs. First, as outlined in our section dedicated to HSCs, there exists significant donor-to-donor heterogeneity in telomere lengths, a feature that is heritable. Donor variations may be so significant that very large cohorts of subjects would be required to generate trends and confidence intervals to establish the best "cut-off-point". Secondly, as outlined in our methodology section, the method used to measure telomeres should be uniform and agreed upon amongst the scientific community, to allow standardisation and adequate comparisons between the manufacturing centres and laboratories.

Extending proliferation span of manufactured MSCs is another attractive possibility. It is a common view that telomere shortening occurs in MSCs due to very low telomerase activity (Fehrer and Lepperdinger 2005). The improved MSC longevity can be achieved via forced hTERT expression in MSCs; however it poses significant dangers relating to tumorogeneicity and is generally not considered as a viable therapeutic option. Culturing MSCs in highly specialised conditions (including fibronectin-coated surfaces and low oxygen tension) can however be considered as an alternative; it yields cell preparations with longer proliferative span and broader potentiality. This could be due to transient re-activation of telomerase or due to alternative telomere lengthening mechanisms. Uncovering such mechanisms would be required in order to bring these *in vitro*-engineered cells to clinic.

The final conclusion from this article relates to the potential use of uncultured, minimally-manipulated MSCs as therapy. It is likely that such cells have the highest proliferative span and best differentiation capability, since they have not been artificially aged in culture. The question however remains as to how many native MSCs can be extracted from individual donors and whether this would give a sufficient dose for therapy. The study of telomere length in rare native cells is currently limited by the lack of reliable purification methods and the amount of material available. Methods requiring a small amount of starting material such as qPCR and flow-FISH may be advantageous, but they need to be further assessed in comparison to more broadly-used assays such as the TRF method. The isolation of native MSCs from mixed cell populations based on phenotypic differences and longer telomeres may be attractive, and the method allowing a direct comparison between putative MSCs and non-MSCs in the same assay, such as flow-FISH, could be the best way forward. This may be most relevant for the study of native MSCs in patients suffering from Dyskeratosis congenital, Werner's syndrome and potentially, osteoporosis. Targeting telomere loss in native MSCs in these patients may represent a novel therapeutic option to treat these diseases in the future.

REFERENCES

Allshire, R. C., Dempster, M. and Hastie, N. D. (1989). "Human telomeres contain at least three types of G–rich repeat distributed non-randomly." N*ucleic Acids Research.* 17(12): 4611-4627.

Allsopp, R. C. and Harley, C. B. (1995). "Evidence for a critical telomere length in senescent human fibroblasts." *Exp. Cell Res.* 219(1): 130 - 136.

Allsopp, R. C., Vaziri, H., Patterson, C. et al. (1992). "Telomere length predicts replicative capacity of human fibroblasts." *Proceedings of the National Academy of Sciences.* 89(21): 10114-10118.

Alt, E., Yan, Y. S., Gehmert, S. et al. (2011). "Fibroblasts share mesenchymal phenotypes with stem cells, but lack their differentiation and colony-forming potential (vol 103, pg 197, 2011)." *Biology of the Cell.* 103(8): 403-403.

Arthur, A., Rychkov, G., Shi, S. et al. (2008). "Adult human dental pulp stem cells differentiate toward functionally active neurons under appropriate environmental cues." *Stem Cells.* 26(7): 1787-1795.

Aubert, G., Hills, M. and Lansdorp, P. M. (2011). "Telomere length measurement—Caveats and a critical assessment of the available technologies and tools." Mutation Research/Fundamental and Molecular Mechanisms of Mutagenesis. 2011 Jun 8 [epub ahead of print]

Aviv, A., Hunt, S. C., Lin, J. et al. (2011). "Impartial comparative analysis of measurement of leukocyte telomere length/DNA content by Southern blots and qPCR." Nucleic Acids Research;39(20):e134.

Baerlocher, G. M., Vulto, I., de Jong, G. et al. (2006). "Flow cytometry and FISH to measure the average length of telomeres (flow FISH)." *Nature Protocols.* 1(5): 2365-2376.

Baird, D. M., Rowson, J., Wynford-Thomas, D. et al. (2003). "Extensive allelic variation and ultrashort telomeres in senescent human cells." *Nature Genetics.* 33(2): 203-207.

Bajada, S., Mazakova, I., Richardson, J. B. et al. (2008). "Updates on stem cells and their applications in regenerative medicine." *Journal of Tissue Engineering and Regenerative Medicine.* 2(4): 169-183.

Baksh, D., Yao, R. and Tuan, R. S. (2007). "Comparison of proliferative and multilineage differentiation potential of human mesenchymal stem cells derived from umbilical cord and bone marrow." Stem Cells. 25(6): 1384-1392.

Banfi, A., Bianchi, G., Notaro, R. et al. (2002). "Replicative aging and gene expression in long-term cultures of human bone marrow stromal cells." *Tissue Engineering.* 8(6): 901-910.

Banfi, A., Muraglia, A., Dozin, B. et al. (2000). "Proliferation kinetics and differentiation potential of ex vivo expanded human bone marrow stromal cells: Implications for their use in cell therapy." *Experimental Hematology.* 28(6): 707-715.

Barry, F. P. and Murphy, J. M. (2004). "Mesenchymal stem cells: clinical applications and biological characterization." *International Journal Of Biochemistry & Cell Biology.* 36(4): 568-584.

Baxter, M. A., Wynn, R. F., Jowitt, S. N. et al. (2004). "Study of telomere length reveals rapid aging of human marrow stromal cells following in vitro expansion." *Stem Cells.* 22(5): 675-682.

Bentzon, J. F., Stenderup, K., Hansen, F. D. et al. (2005). "Tissue distribution and engraftment of human mesenchymal stem cells immortalized by human telomerase reverse transcriptase gene." *Biochemical and Biophysical Research Communications.* 330(3): 633-640.

Bernardo, M. E., Zaffaroni, N., Novara, F. et al. (2007). "Human bone marrow-derived mesenchymal stem cells do not undergo transformation after long-term In vitro culture and do not exhibit telomere maintenance mechanisms." *Cancer Research.* 67(19): 9142-9149.

Bianchi, G., Banfi, A., Mastrogiacomo, M. et al. (2003). "Ex vivo enrichment of mesenchymal cell progenitors by fibroblast growth factor 2." *Experimental Cell Research.* 287(1): 98-105.

Bianco, P., Robey, P. G., Saggio, I. et al. (2010). "Mesenchymal" stem cells in human bone marrow (skeletal stem cells): A critical discussion of their nature, identity, and significance in incurable skeletal disease." *Human Gene Therapy.* 21(9): 1057-1066.

Bonab, M. M., Alimoghaddam, K., Talebian, F. et al. (2006). "Aging of mesenchymal stem cell in vitro." *BMC Cell Biology.* 7(14).

Bruder, S. P., Jaiswal, N. and Haynesworth, S. E. (1997). "Growth kinetics, self-renewal, and the osteogenic potential of purified human mesenchymal stem cells during extensive subcultivation and following cryopreservation." *Journal of Cellular Biochemistry.* 64(2): 278-294.

Bryant, J. E., Hutchings, K. G., Moyzis, R. K. et al. (1997). "Measurement of telomeric DNA content in human tissues." *Biotechniques.* 23(3): 476-478.

Buhring, H.-J., Battula, V. L., Treml, S. et al. (2007). "Novel markers for the prospective isolation of human MSC." *Ann. NY Acad. Sci.* 1106: 262-271.

Buhring, H. J., Treml, S., Cerabona, F. et al. (2009). Phenotypic characterization of distinct human bone marrow-derived MSC subsets. Hematopoietic stem cells Vii. L. Kanz, K. C. Weisel, J. E. Dick and W. E. Fibbe. Oxford, Blackwell Publishing. 1176: 124-134.

Burns, J. S., Abdallah, B. M., Guldberg, P. et al. (2005). "Tumorigenic heterogeneity in cancer stem cells evolved from long-term cultures of telomerase-immortalized human mesenchymal stem cells." *Cancer Research.* 65(8): 3126-3135.

Campard, D., Lysy, P. A., Najimi, M. et al. (2008). "Native umbilical cord matrix stem cells express hepatic markers and differentiate into hepatocyte-like cells." *Gastroenterology.* 134(3): 833-848.

Can, A. and Karahuseyinoglu, S. (2007). "Concise review: Human umbilical cord stroma with regard to the rource of fetus-derived stem cells." *Stem Cells.* 25(11): 2886-2895.

Canela, A., Vera, E., Klatt, P. et al. (2007). "High-throughput telomere length quantification by FISH and its application to human population studies." *Proceedings of the National Academy of Sciences of the United States of America.* 104(13): 5300-5305.

Carbonari, M., Mancaniello, D., Cibati, M. et al. (2010). "Improved procedure for the measurement of telomere length in whole cells by PNA probe and flow cytometry." *Cell Proliferation.* 43(6): 553-561.

Carlin, R., Davis, D., Weiss, M. et al. (2006). "Expression of early transcription factors Oct-4, Sox-2 and Nanog by porcine umbilical cord (PUC) matrix cells." *Reproductive Biology and Endocrinology.* 4(1): 8.

Cassidy, A., De Vivo, I., Liu, Y. et al. (2010). "Associations between diet, lifestyle factors, and telomere length in women." *American Journal of Clinical Nutrition.* 91(5): 1273-1280.

Cawthon, R. M. (2002). "Telomere measurement by quantitative PCR." *Nucleic Acids Research.* 30(10): e47.

Cawthon, R. M. (2009). "Telomere length measurement by a novel monochrome multiplex quantitative PCR method." *Nucleic Acids Research.* 37(3): e21.

Cherkas, L. F., Hunkin, J. L., Kato, B. S. et al. (2008). "The association between physical activity in leisure time and leukocyte telomere length." *Arch. Intern. Med.* 168(2): 154-158.

Civin, C. I., Trischmann, T., Kadan, N. S. et al. (1996). "Highly purified CD34-positive cells reconstitute hematopoiesis." *Journal of Clinical Oncology.* 14(8): 2224-2233.

Cohen, Y. and Nagler, A. (2004). *Cord blood biology and transplantation.* 6(1): 39-46

Cong, Y.-S., Wright, W. E. and Shay, J. W. (2002). "Human telomerase and its regulation." *Microbiol. Mol. Biol. Rev.* 66(3): 407-425.

Cousin, B. a., AndrΓ©, M., Arnaud, E. et al. (2003). "Reconstitution of lethally irradiated mice by cells isolated from adipose tissue." *Biochemical and Biophysical Research Communications.* 301(4): 1016-1022.

Covas, D. T., Panepucci, R. A., Fontes, A. M. et al. (2008). "Multipotent mesenchymal stromal cells obtained from diverse human tissues share functional properties and gene-expression profile with CD146(+) perivascular cells and fibroblasts." *Experimental Hematology.* 36(5): 642-654.

Cowan, C. M., Shi, Y.-Y., Aalami, O. O. et al. (2004). "Adipose-derived adult stromal cells heal critical-size mouse calvarial defects." *Nat. Biotech.* 22(5): 560-567.

Crisan, M., Yap, S., Casteilla, L. et al. (2008). "A perivascular origin for mesenchymal stem cells in multiple human organs." *Cell Stem Cell.* 3(3): 301-313.

d'Aquino, R., Graziano, A., Sampaolesi, M. et al. (2007). "Human postnatal dental pulp cells co-differentiate into osteoblasts and endotheliocytes: a pivotal synergy leading to adult bone tissue formation." *Cell Death Differ.* 14(6): 1162-1171.

D'Ippolito, G., Diabira, S., Howard, G. A. et al. (2004). "Marrow-isolated adult multilineage inducible (MIAMI) cells, a unique population of postnatal young and old human cells with extensive expansion and differentiation potential." *Journal of Cell Science.* 117(14): 2971-2981.

De Bari, C., Dell'Accio, F., Karystinou, A. et al. (2008). "A biomarker-based mathematical model to predict bone-forming potency of human synovial and periosteal mesenchymal stem cells." *Arthritis & Rheumatism.* 58(1): 240.

De Bari, C., Dell'Accio, F., Tylzanowski, P. et al. (2001). "Multipotent mesenchymal stem cells from adult human synovial membrane." *Arthritis & Rheumatism.* 44(8): 1928.

De Bari, C., Dell'Accio, F., Vanlauwe, J. et al. (2006). "Mesenchymal multipotency of adult human periosteal cells demonstrated by single-cell lineage analysis." *Arthritis & Rheumatism.* 54(4): 1209-1221.

Dempster, D. W. (2002). "The impact of bone turnover and bone-active agents on bone quality: Focus on the hip." *Osteoporosis International.* 13(5): 349-352.

Dominici, M., Le Blanc, K., Mueller, I. et al. (2006). "Minimal criteria for defining multipotent mesenchymal stromal cells. The International Society for Cellular Therapy position statement." *Cytotherapy.* 8(4): 315-317.

Efimenko, A., Starostina, E., Kalinina, N. et al. (2011). "Angiogenic properties of aged adipose derived mesenchymal stem cells after hypoxic conditioning." *Journal of Translational Medicine.* 9(10).

Elshal, M. F., Khan, S. S., Takahashi, Y. et al. (2005). "CD146 (Mel-CAM), an adhesion marker of endothelial cells, is a novel marker of lymphocyte subset activation in normal peripheral blood." *Blood.* 106(8): 2923-2924.

Endres, M., Andreas, K., Kalwitz, G. et al. (2010). "Chemokine profile of synovial fluid from normal, osteoarthritis and rheumatoid arthritis patients: CCL25, CXCL10 and XCL1 recruit human subchondral mesenchymal progenitor cells." *Osteoarthritis and Cartilage.* 18(11): 1458-1466.

Epel, E. S., Blackburn, E. H., Lin, J. et al. (2004). "Accelerated telomere shortening in response to life stress." *Proceedings of the National Academy of Sciences of the United States of America.* 101(49): 17312-17315.

Erices, A., Conget, P. and Minguell, J. J. (2000). "Mesenchymal progenitor cells in human umbilical cord blood." *British Journal of Haematology.* 109(1): 235.

Fan, C.-G., Zhang, Q.-j. and Zhou, J.-r. (2011). "Therapeutic potentials of mesenchymal stem cells derived from human umbilical cord." *Stem Cell Reviews and Reports.* 7(1): 195-207.

Fehrer, C. and Lepperdinger, G. (2005). "Mesenchymal stem cell aging." *Experimental Gerontology.* 40(12): 926.

Flores, I., Canela, A., Vera, E. et al. (2008). "The longest telomeres: a general signature of adult stem cell compartments." *Genes & Development.* 22(5): 654-667.

Fong, C.-Y., Subramanian, A., Biswas, A. et al. (2010). "Derivation efficiency, cell proliferation, freeze-thaw survival, stem-cell properties and differentiation of human Wharton's jelly stem cells." *Reproductive Biomedicine Online.* 21(3): 391-401.

Forsyth, N. R. and McWhir, J. (2008). "Human embryonic stem cell telomere length impacts directly on clonal progenitor isolation frequency." *Rejuvenation Research.* 11(1): 5-17.

Foster, L. J., Zeemann, P. A., Li, C. et al. (2005). "Differential expression profiling of membrane proteins by quantitative proteomics in a human mesenchymal stem cell line undergoing osteoblast differentiation." *Stem Cells.* 23(9): 1367.

Friedenstein, A. J., Chailakhyan, R. K. and Gerasimov, U. V. (1987). "Bone-Marrow osteogenic stem cells - In vitro cultivation and transplantation in diffusion chambers." *Cell And Tissue Kinetics.* 20(3): 263-272.

Friedenstein, A. J., Latzinik, N. V., Gorskaya, Y. F. et al. (1992). "Bone-marrow stromal colony formation requires stimulation by hematopoietic cells." *Bone and Mineral.* 18(3): 199-213.

Friedenstein, A. J., Latzinik, N. W., Grosheva, A. G. et al. (1982). "Marrow microenvironment transfer by heterotopic transplantation of freshly isolated and cultured cells in porous sponges." *Exp. Hematol.* 10(2): 217 - 227.

Friedrich, U., Griese, E.-U., Schwab, M. et al. (2000). "Telomere length in different tissues of elderly patients." *Mechanisms of Ageing and Development.* 119(3): 89-99.

Fukumoto, T., Sperling, J. W., Sanyal, A. et al. (2003). "Combined effects of insulin-like growth factor-1 and transforming growth factor-b1 on periosteal mesenchymal cells during chondrogenesis in vitro." *Osteoarthritis and Cartilage.* 11(1): 55-64.

Gadalla, S. M., Cawthon, R., Giri, N. et al. (2010). "Telomere length in blood, buccal cells, and fibroblasts from patients with inherited bone marrow failure syndromes." *Aging.* 2(11): 867-874.

Gadalla, S. M., Katki, H. A., Shebl, F. M. et al. (2011). "The relationship between DNA methylation and telomere length in dyskeratosis congenita." *Aging Cell.* 2(11): 867-874.

Gandia, C., Armiñan, A., García-Verdugo, J. M. et al. (2008). "Human dental pulp stem cells improve left ventricular function, induce angiogenesis, and reduce infarct size in rats with acute myocardial infarction." *Stem Cells.* 26(3): 638-645.

Gang, E. J., Bosnakovski, D., Figueiredo, C. A. et al. (2007). "SSEA-4 identifies mesenchymal stem cells from bone marrow." *Blood.* 109(4): 1743-1751.

Gertler, R., Rosenberg, R., Stricker, D. et al. (2004). "Telomere length and human telomerase reverse transcriptase expression as markers for progression and prognosis of colorectal carcinoma." *Journal of Clinical Oncology.* 22(10): 1807-1814.

Glauche, I., Thielecke, L. and Roeder, I. (2011). "Cellular aging leads to functional heterogeneity of hematopoietic stem cells: a modeling perspective." *Aging Cell.* 10(3): 457-465.

Goodwin, H. S., Bicknese, A. R., Chien, S. N. et al. (2001). "Multilineage differentiation activity by cells isolated from umbilical cord blood: Expression of bone, fat, and neural markers." *Biology of blood and marrow transplantation.* 7(11): 581-588.

Goronzy, J. and Weyand, C. (2003). "Ageing, autoimmunity and arthritis: T-cell senescence and contraction of T-cell repertoire diversity - catalysts of autoimmunity and chronic inflammation." *Arthritis Res. Ther.* 5(5): 225 - 234.

Goto, M. (2000). "Werner's syndrome: From clinics to genetics." *Clinical and Experimental Rheumatology.* 18(6): 760-766.

Gregory, C. A., Ylostalo, J. and Prockop, D. J. (2005). "Adult bone marrow stem/progenitor cells (MSCs) are preconditioned by microenvironmental "niches" in culture: A two-stage hypothesis for regulation of MSC fate." *Sci. STKE.* 2005(294): pe37.

Gronthos, S., Brahim, J., Li, W. et al. (2002). "Stem cell properties of human dental pulp stem cells."*Journal of Dental Research.* 81(8): 531-535.

Guillot, P. V., De Bari, C., Dell'Accio, F. et al. (2008). "Comparative osteogenic transcription profiling of various fetal and adult mesenchymal stem cell sources." *Differentiation.* 76(9): 946-957.

Guillot, P. V., Gotherstrom, C., Chan, J. et al. (2007). "Human first-trimester fetal MSC express pluripotency markers and grow faster and have longer telomeres than adult MSC." *Stem Cells.* 25(3): 646-654.

Gunsilius, E., Gastl, G. and Petzer, A. L. (2001). "Hematopoietic stem cells." *Biomedicine & Pharmacotherapy.* 55(4): 186-194.

Halaschek-Wiener, J., Vulto, I., Fornika, D. et al. (2008). "Reduced telomere length variation in healthy oldest old." *Mechanisms of Ageing and Development.* 129(11): 638-641.

Helder, M. N., Knippenberg, M., Klein-Nulend, J. et al. (2007). "Stem Cells from Adipose Tissue Allow Challenging New Concepts for Regenerative Medicine." *Tissue Engineering.* 13(8): 1799-1808.

Hills, M., Luecke, K., Chavez, E. A. et al. (2009). "Probing the mitotic history and developmental stage of hematopoietic cells using single telomere length analysis (STELA)." *Blood.* 113(23): 5765-5775.

Huang, G. T. J., Gronthos, S. and Shi, S. (2009). "Mesenchymal stem cells derived from dental tissues vs. those from other sources: their biology and role in regenerative medicine." *Journal of Dental Research.* 88(9): 792-806.

Iohara, K., Nakashima, M., Ito, M. et al. (2004). "Dentin regeneration by dental pulp stem cell therapy with recombinant human bone morphogenetic protein 2." *Journal of Dental Research.* 83(8): 590-595.

Ishige, I., Nagamura-Inoue, T., Honda, M. et al. (2009). "Comparison of mesenchymal stem cells derived from arterial, venous, and Wharton's jelly explants of human umbilical cord." *International Journal of Hematology.* 90(2): 261-269.

Ishii, M., Koike, C., Igarashi, A. et al. (2005). "Molecular markers distinguish bone marrow mesenchymal stem cells from fibroblasts." *Biochemical and Biophysical Research Communications.* 332(1): 297-303.

Ishikawa, N., Nakamura, K., Izumiyama-Shimomura, N. et al. (2011). "Accelerated in vivo epidermal telomere loss in Werner syndrome." *Aging.* 3(4): 417-429.

Izadpanah, R., Trygg, C., Patel, B. et al. (2006). "Biologic properties of mesenchymal stem cells derived from bone marrow and adipose tissue." *Journal of Cellular Biochemistry.* 99(5): 1285-1297.

Jeanclos, E., Schork, N. J., Kyvik, K. O. et al. (2000). "Telomere length inversely correlates with pulse pressure and is highly familial." *Hypertension.* 36(2): 195-200.

Jeon, B.-G., Kumar, B., Kang, E.-J. et al. (2011). "Characterization and comparison of telomere length, telomerase and reverse transcriptase activity and gene expression in human mesenchymal stem cells and cancer cells of various origins." *Cell and Tissue Research.* 345(1): 149-161.

Jo, C., Kim, O.-S., Park, E.-Y. et al. (2008). "Fetal mesenchymal stem cells derived from human umbilical cord sustain primitive characteristics during extensive expansion." *Cell and Tissue Research.* 334(3): 423-433.

Jones, E., English, A., Churchman, S. M. et al. (2010). "Large-scale extraction and characterization of CD271+ multipotential stromal cells from trabecular bone in health and osteoarthritis: implications for bone regeneration strategies based on uncultured or minimally cultured multipotential stromal cells." *Arthritis and Rheumatism* 62(7): 1944-1954.

Jones, E. and McGonagle, D. (2008). "Human bone marrow mesenchymal stem cells in vivo." *Rheumatology.* 47(2): 126-131.

Jones, E. and Yang, X. B. (2011). "Mesenchymal stem cells and bone regeneration: Current status." *Injury.* 42(6): 562-568.

Jones, E. A., Crawford, A., English, A. et al. (2008). "Synovial fluid mesenchymal stem cells in health and early osteoarthritis: Detection and functional evaluation at the single-cell level." *Arthritis and Rheumatism* 58(6): 1731-1740.

Jones, E. A., English, A., Emery, P. et al. (2002). "Isolation and characterization of unique human bone marrow multipotential mesenchymal progenitor cells." *Arthritis and Rheumatism* 46(9): S568-S568.

Jones, E. A., English, A., Henshaw, K. et al. (2004). "Enumeration and phenotypic characterization of synovial fluid multipotential mesenchymal progenitor cells in inflammatory and degenerative arthritis." *Arthritis and Rheumatism.* 50(3): 817-827.

Kamminga, L. M., Bystrykh, L. V., de Boer, A. et al. (2006). "The Polycomb group gene Ezh2 prevents hematopoietic stem cell exhaustion." *Blood.* 107(5): 2170-2179.

Kang, S. K., Lee, D. H., Bae, Y. C. et al. (2003). "Improvement of neurological deficits by intracerebral transplantation of human adipose tissue-derived stromal cells after cerebral ischemia in rats." *Experimental Neurology.* 183(2): 355-366.

Kang, S. K., Putnam, L., Dufour, J. et al. (2004). "Expression of telomerase extends the lifespan and enhances osteogenic differentiation of adipose tissue-derived stromal cells." *Stem Cells.* 22(7): 1356-1372.

Karahuseyinoglu, S., Cinar, O., Kilic, E. et al. (2007). "Biology of Stem Cells in Human Umbilical Cord Stroma: In Situ and In Vitro Surveys." *Stem Cells.* 25(2): 319-331.

Kastrinaki, M. C., Spanoudakis, M., Sidiropoulos, P. et al. (2006). "Bone marrow mesenchymal stem cell properties in patients with rheumatoid arthritis." *Annals of the Rheumatic Diseases.* 65: A37-A38.

Kim, J., Kang, J. W., Park, J. H. et al. (2009). "Biological characterization of long-term cultured human mesenchymal stem cells." *Archives of Pharmacal Research.* 32(1): 117-126.

Kimura, M. and Aviv, A. (2011). "Measurement of telomere DNA content by dot blot analysis." *Nucleic Acids Research.* 39(12): e84.

Kimura, M., Gazitt, Y., Cao, X. et al. (2010). "Synchrony of telomere length among hematopoietic cells." *Experimental Hematology.* 38(10): 854-859.

Kögler, G., Sensken, S., Airey, J. A. et al. (2004). "A new human somatic stem cell from placental cord blood with intrinsic pluripotent differentiation potential." *The Journal of Experimental Medicine.* 200(2): 123-135.

Kurth, T. B., Dell'Accio, F., Crouch, V. et al. (2011). "Functional mesenchymal stem cell nches in adult mouse knee joint synovium in vivo." *Arthritis and Rheumatism.* 63(5): 1289-1300.

La Rocca, G., Anzalone, R., Corrao, S. et al. (2009). "Isolation and characterization of Oct-4+/HLA-G+ mesenchymal stem cells from human umbilical cord matrix: differentiation potential and detection of new markers." *Histochemistry and Cell Biology.* 131(2): 267-282.

Laino, G., D'Aquino, R., Graziano, A. et al. (2005). "A new population of human adult dental pulp stem cells: A useful source of living autologous fibrous bone tissue (LAB)." *Journal of Bone and Mineral Research.* 20(8): 1394-1402.

Lansdorp, P. M., Verwoerd, N. P., van de Rijke, F. M. et al. (1996). "Heterogeneity in telomere length of human chromosomes." *Human Molecular Genetics.* 5(5): 685-691.

Le Blanc, K., Rasmusson, I., Sundberg, B. et al. (2004). "Treatment of severe acute graft-versus-host disease with third party haploidentical mesenchymal stem cells." *The Lancet.* 363(9419): 1439-1441.

Le Blanc, K., Tammik, L., Sundberg, B. et al. (2003). "Mesenchymal stem cells inhibit and stimulate mixed lymphocyte cultures and mitogenic responses independently of the major histocompatibility complex." *Scandinavian Journal of Immunology.* 57(1): 11-20.

Lee, C. H., Cook, J. L., Mendelson, A. et al. (2010). "Regeneration of the articular surface of the rabbit synovial joint by cell homing: a proof of concept study." *Lancet.* 376(9739): 440-448.

Levi, B., James, A. W., Glotzbach, J. P. et al. (2010). "Depot-specific variation in the osteogenic and adipogenic potential of human adipose-derived stromal cells." *Plastic and Reconstructive Surgery.* 126(3): 822-834.

Mackay, A. M., Beck, S. C., Murphy, J. M. et al. (1998). "Chondrogenic differentiation of cultured human mesenchymal stem cells from marrow." *Tissue Engineering.* 4(4): 415-428.

Majore, I., Moretti, P., Stahl, F. et al. (2011). "Growth and differentiation properties of mesenchymal stromal cell populations derived from whole human umbilical cord." *Stem Cell Reviews and Reports.* 7(1): 17-31.

Mareschi, K., Ferrero, I., Rustichelli, D. et al. (2006). "Expansion of mesenchymal stem cells isolated from pediatric and adult donor bone marrow." *Journal of Cellular Biochemistry.* 97(4): 744-754.

Marrone, A., Stevens, D., Vulliamy, T. et al. (2004). "Heterozygous telomerase RNA mutations found in dyskeratosis congenita and aplastic anemia reduce telomerase activity via haploinsufficiency." *Blood.* 104(13): 3936-3942.

Martinez, C., Hofmann, T. J., Marino, R. et al. (2007). "Human bone marrow mesenchymal stromal cells express the neural ganglioside GD2: a novel surface marker for the identification of MSCs." *Blood.* 109(10): 4245-4248.

McBeath, R., Pirone, D. M., Nelson, C. M. et al. (2004). "Cell shape, cytoskeletal tension, and RhoA regulate stem cell lineage commitment." *Developmental Cell.* 6(4): 483-495.

Mehrazarin, S., Oh, J. E., Chung, C. L. et al. (2011) "Impaired odontogenic differentiation of senescent dental mesenchymal stem cells is associated with loss of Bmi-1 expression." *Journal of Endodontics.* 37(5): 662-666.

Meirelles, L. d. S. and Nardi, N. B. (2003). Murine marrow-derived mesenchymal stem cell: isolation, in vitro expansion, and characterization. *Br. J. Haematol.* 123(4): 702-711.

Mendelson, A., Frank, E., Allred, C. et al. (2011). "Chondrogenesis by chemotactic homing of synovium, bone marrow, and adipose stem cells in vitro." *The FASEB journal.* 25(10): 3496-3504.

Mirsaidi, A., Kleinhans, K. N., Rimann, M. et al. (2011). "Telomere length, telomerase activity and osteogenic differentiation are maintained in adipose-derived stromal cells from senile osteoporotic SAMP6 mice." *Journal of Tissue Engineering and Regenerative Medicine.* June 28 [epub ahead of print].

Mitchell, J. R., Wood, E. and Collins, K. (1999). "A telomerase component is defective in the human disease dyskeratosis congenita." *Nature.* 402(6761): 551-555.

Miura, M., Gronthos, S., Zhao, M. et al. (2003). "SHED - Stem cells from human exfoliated deciduous teeth." *Journal of Dental Research.* 82: B305-B305.

Mizuno, H. (2009). "Adipose-derived stem cells for tissue repair and regeneration: Ten years of research and a literature review." *Journal of Nippon Medical School.* 76(2): 56-66.

Mokry, J., Soukup, T., Micuda, S. et al. (2010). "Telomere attrition occurs during ex vivo expansion of human dental pulp stem cells." *Journal of Biomedicine and Biotechnology.* 673513.

Moretti, P., Hatlapatka, T., Marten, D. et al. (2010) "Mesenchymal stromal cells derived from human umbilical cord tissues: primitive cells with potential for clinical and tissue engineering applications." *Adv. Biochem. Engin. Biotechnol.* 123: 29-54.

Morsczeck, C., Götz, W., Schierholz, J. et al. (2005). "Isolation of precursor cells (PCs) from human dental follicle of wisdom teeth." *Matrix Biology.* 24(2): 155-165.

Moyzis, R. K., Buckingham, J. M., Cram, L. S. et al. (1988). "A highly conserved repetitive DNA-Sequence, (Ttaggg)N, Present at the telomeres of human chromosomes." *Proceedings of the National Academy of Sciences of the United States of America* 85(18): 6622-6626.

Muraglia, A., Cancedda, R. and Quarto, R. (2000). "Clonal mesenchymal progenitors from human bone marrow differentiate in vitro according to a hierarchical model." *Journal of Cell Science.* 113(7): 1161-1166.

Nawrot, T. S., Staessen, J. A., Gardner, J. P. et al. (2004). "Telomere length and possible link to X chromosome." *The Lancet.* 363(9408): 507-510.

Nelson, N. D. and Bertuch, A. A. (2011) "Dyskeratosis congenita as a disorder of telomere maintenance." Mutation Research. July 12 [epub ahead of print].

Noer, A., Boquest, A. and Collas, P. (2007). "Dynamics of adipogenic promoter DNA methylation during clonal culture of human adipose stem cells to senescence." *BMC Cell Biology.* 8(1): 18-28.

Nordfjall, K., Svenson, U., Norrback, K.-F. et al. (2009). "Large-scale parent-child comparison confirms a strong paternal influence on telomere length." *Eur. J. Hum. Genet.* 18(3): 385-389.

Oeseburg, H., de Boer, R. A., van Gilst, W. H. et al. (2010). "Telomere biology in healthy aging and disease." *Pflugers Archiv-European Journal of Physiology*. 459(2): 259-268.

Ogden, D. A. and Micklem, H. S. (1976). "Fate of serially transplanted bone-Marrow cell-populations from young and old donors." *Transplantation.* 22(3): 287-293.

Oktar, P., Yildirim, S., Balci, D. et al. (2011). "Continual expression throughout the cell cycle and downregulation upon adipogenic differentiation makes nucleostemin a vital human MSC proliferation marker." *Stem Cell Reviews and Reports.* 7(2): 413-424.

Pacini, S., Carnicelli, V., Trombi, L. et al. (2010). "Constitutive expression of pluripotency-associated genes in mesodermal progenitor cells (MPCs)." *PLoS ONE.* 5(3): e9861.

Parfitt, A. (1984). "The cellular basis of bone remodeling: The quantum concept reexamined in light of recent advances in the cell biology of bone." *Calcified Tissue International.* 36(0): S37-S45.

Parfitt, A. M. (2002). "Misconceptions (2): Turnover is always higher in cancellous than in cortical bone." *Bone.* 30(6): 807-809.

Pasquinelli, G., Tazzari, P., Ricci, F. et al. (2007). "Ultrastructural characteristics of human mesenchymal stromal (stem) cells derived from bone marrow and term placenta." *Ultrastructural Pathology.* 31(1): 23 - 31.

Piacibello, W., Gammaitoni, L. and Pignochino, Y. (2005). "Proliferative senescence in hematopoietic stem cells during ex-vivo expansion." *Folia Histochemica Et Cytobiologica.* 43(4): 197-202.

Pignolo, R. J., Suda, R. K., McMillan, E. A. et al. (2008). "Defects in telomere maintenance molecules impair osteoblast differentiation and promote osteoporosis." *Aging Cell.* 7(1): 23-31.

Pittenger, M. F., Mackay, A. M., Beck, S. C. et al. (1999). "Multilineage potential of adult human mesenchymal stem cells." *Science.* 284(5411): 143-147.

Poon, S. S. S. and Lansdorp, P. M. (2001). Quantitative fluorescence in situ hybridization (Q-FISH). *Current Protocols in Cell Biology chapter.* 18:unit 18.4.

Price, J. S., Waters, J. G., Darrah, C. et al. (2002). "The role of chondrocyte senescence in osteoarthritis." *Aging Cell.* 1(1): 57-65.

Prockop, D. J. (2009). "Repair of tissues by adult stem/progenitor cells (MSCs): Controversies, myths, and changing paradigms." *Molecular Therapy.* 17(6): 939-946.

Quarto, R., Mastrogiacomo, M., Cancedda, R. et al. (2001). "Repair of large bone defects with the use of autologous bone marrow stromal cells." *New England Journal of Medicine.* 344(5): 385-386.

Quirici, N., Scavullo, C., de Girolamo, L. et al. (2010). "Anti-L-NGFR and-CD34 Monoclonal antibodies identify multipotent mesenchymal stem cells in human adipose tissue." *Stem Cells and Development.* 19(6): 915-925.

Rebuzzini, P., Neri, T., Mazzini, G. et al. (2008). "Karyotype analysis of the euploid cell population of a mouse embryonic stem cell line revealed a high incidence of chromosome abnormalities that varied during culture." *Cytogenetic and Genome Research.* 121(1): 18-24.

Reyes, M., Lund, T., Lenvik, T. et al. (2001). "Purification and ex vivo expansion of postnatal human marrow mesodermal progenitor cells." *Blood.* 98(9): 2615-2625.

Rufer, N., BrΓOmmendorf, T. H., Kolvraa, S. et al. (1999). "Telomere fluorescence measurements in granulocytes and T lymphocyte subsets point to a high turnover of hematopoietic stem cells and memory T cells in early childhood." *The Journal of Experimental Medicine.* 190(2): 157-168.

Rufer, N., Dragowska, W., Thornbury, G. et al. (1998). "Telomere length dynamics in human lymphocyte subpopulations measured by flow cytometry." *Nat. Biotech.* 16(8): 743-747.

Sakaguchi, Y., Sekiya, I., Yagishita, K. et al. (2005). "Comparison of human stem cells derived from various mesenchymal tissues - Superiority of synovium as a cell source." *Arthritis and Rheumatism.* 52(8): 2521-2529.

Sanders, J. L., Cauley, J. A., Boudreau, R. M. et al. (2009). "Leukocyte telomere length is not associated with BMD, osteoporosis, or fracture in older adults: Results from the Health, Aging and Body Composition Study." *Journal of Bone and Mineral Research.* 24(9): 1531-1536.

Schmid, I., Dagarag, M. D., Hausner, M. A. et al. (2002). "Simultaneous flow cytometric analysis of two cell surface markers, telomere length, and DNA content." *Cytometry.* 49(3): 96-105.

Schugar, R. C., Chirieleison, S. M., Wescoe, K. E. et al. (2009). "High harvest yield, high expansion, and phenotype stability of CD146 mesenchymal stromal cells from whole primitive human umbilical cord tissue." *Journal of Biomedicine and Biotechnology*. doi: 10.1155/2009/789526.

Sekiya, I., Larson, B. L., Smith, J. R. et al. (2002). "Expansion of human adult stem cells from bone marrow stroma: Conditions that maximize the yields of early progenitors and evaluate their quality." *Stem Cells*. 20(6): 530-541.

Seo, B. M., Miura, M., Gronthos, S. et al. (2004). "Investigation of multipotent postnatal stem cells from human periodontal ligament." *Lancet*. 364(9429): 149-155.

Seong, J. M., Kim, B.-C., Park, J.-H. et al. (2010). "Stem cells in bone tissue engineering." *Biomedical Materials* (6):062001.

Sessarego, N., Parod, A., Podesta, M. et al. (2008). "Multipotent mesenchymal stromal cells from amniotic fluid: solid perspectives for clinical application." *Haematologica*. 93(3): 339-346.

Sidorov, I., Kimura, M., Yashin, A. et al. (2009). "Leukocyte telomere dynamics and human hematopoietic stem cell kinetics during somatic growth." *Experimental Hematology*. 37(4): 514-524.

Simonsen, J. L., Rosada, C., Serakinci, N. et al. (2002). "Telomerase expression extends the proliferative life-span and maintains the osteogenic potential of human bone marrow stromal cells." *Nat. Biotech*. 20(6): 592-596.

Slagboom, P. E., Droog, S. and Boomsma, D. I. (1994). "Genetic determination of telomere size in humans: a twin study of three age groups." *American journal of human genetics*. 55(5): 876-882.

Sonoyama, W., Liu, Y., Fang, D. et al. (2006). "Mesenchymal stem cell-mediated functional tooth regeneration in swine." *PLoS ONE*. 1(1): e79.

Sotiropoulou, P. A., Perez, S. A., Salagianni, M. et al. (2006). "Characterization of the optimal culture conditions for clinical scale production of human mesenchymal stem cells." *Stem Cells*. 24(2): 462-471.

Stenderup, K., Justesen, J., Clausen, C. et al. (2003). "Aging is associated with decreased maximal life span and accelerated senescence of bone marrow stromal cells." *Bone*. 33(6): 919 - 926.

Stolzing, A., Coleman, N. and Scutt, A. (2006). "Glucose-induced replicative senescence in mesenchymal stem cells." *Rejuvenation Research*. 9(1): 31-35.

Stolzing, A., Jones, E., McGonagle, D. et al. (2008). "Age-related changes in human bone marrow-derived mesenchymal stem cells: Consequences for cell therapies." *Mechanisms of Ageing and Development*. 129(3): 163-173.

Sudo, K., Kanno, M., Miharada, K. et al. (2007). "Mesenchymal progenitors able to differentiate into osteogenic, chondrogenic, and/or adipogenic cells in vitro are present in most primary fibroblast-like cell populations." *Stem Cells*. 25(7): 1610-1617.

Tahara, H., Kusunoki, M., Yamanaka, Y. et al. (2005). "G-tail telomere HPA: simple measurement of human single-stranded telomeric overhangs." *Nature Methods*. 2(11): 829-831.

Takubo, K., Izumiyama-Shimomura, N., Honma, N. et al. (2002). "Telomere lengths are characteristic in each human individual." *Experimental Gerontology*. 37(4): 523-531.

Tamayo, M., Mosquera, A., Ignacio Rego, J. et al. (2010). "Differing patterns of peripheral blood leukocyte telomere length in rheumatologic diseases." *Mutation Research-Fundamental and Molecular Mechanisms of Mutagenesis.* 683(1-2): 68-73.

Tarte, K., Gaillard, J., Lataillade, J. J. et al. (2010). "Clinical-grade production of human mesenchymal stromal cells: occurrence of aneuploidy without transformation." *Blood.* 115(8): 1549-1553.

Terai, M., Uyama, T., Sugiki, T. et al. (2005). "Immortalization of human fetal cells: The life span of umbilical cord blood-derived cells can be prolonged without manipulating p16INK4a/RB Braking Pathway." *Molecular Biology of the Cell.* 16(3): 1491-1499.

Tormin, A., Li, O., Brune, J. C. et al. (2011). "CD146 expression on primary non-hematopoietic bone marrow stem cells correlates to in situ localization." *Blood.* 117(19): 5067-5077.

Valdes, A. M., Richards, J. B., Gardner, J. P. et al. (2007). "Telomere length in leukocytes correlates with bone mineral density and is shorter in women with osteoporosis." *Osteoporosis International.* 18(9): 1203-1210.

Vaziri, H., Dragowska, W., Allsopp, R. C. et al. (1994). "Evidence for a mitotic clock in human hematopoietic stem-cells - Loss of telomeric DNA with age." *Proceedings of the National Academy of Sciences of the United States of America.* 91(21): 9857-9860.

Vidal, M. A., Walker, N. J., Napoli, E. et al. (2011). "Evaluation of senescence in mesenchymal stem cells isolated from equine bone marrow, adipose tissue, and umbilical cord tissue." *Stem Cells and Development.* May 6 [epub ahead of print].

Wagner, W., Bork, S., Horn, P. et al. (2009). "Aging and replicative senescence have related effects on human stem and progenitor cells." *PLoS ONE.* 4(6): e5846.

Wagner, W., Horn, P., Castoldi, M. et al. (2008). "Replicative senescence of mesenchymal stem cells: A continuous and organized process." *PLoS ONE.* 3(5): e2213.

Walne, A., Marrone, A. and Dokal, I. (2005). "Dyskeratosis congenita: A disorder of defective telomere maintenance?" *International Journal of Hematology.* 82(3): 184-189.

Wang, H. S., Hung, S. C., Peng, S. T. et al. (2004). "Mesenchymal stem cells in the Wharton's jelly of the human umbilical cord." *Stem Cells.* 22(7): 1330-1337.

Weiss, M. L., Medicetty, S., Bledsoe, A. R. et al. (2006). "Human umbilical cord matrix stem cells: Preliminary characterization and effect of transplantation in a rodent model of Parkinson's disease." *Stem Cells.* 24(3): 781-792.

Wolbank, S., Stadler, G., Peterbauer, A. et al. (2009). "Telomerase immortalized human amnion- and adipose-derived mesenchymal stem cells: maintenance of differentiation and immunomodulatory characteristics." *Tissue Engineering Part A.* 15(7): 1843-1854.

Wu, L.-F., Wang, N.-N., Liu, Y.-S. et al. (2009). "Differentiation of wharton's jelly primitive stromal cells into insulin-producing cells in comparison with bone marrow mesenchymal stem cells." *Tissue Engineering Part A.* 15(10): 2865-2873.

Yamaguchi, H. (2007). "Mutations of telomerase complex genes linked to bone marrow failures." *Journal of Nippon Medical Scool.* 74(3): 202-209.

Yanada, S., Ochi, M., Kojima, K. et al. (2006). "Possibility of selection of chondrogenic progenitor cells by telomere length in FGF-2-expanded mesenchymal stromal cells." *Cell Proliferation.* 39(6): 575-584.

Yang, C.-C., Shih, Y.-H., Ko, M.-H. et al. (2008). "Transplantation of human umbilical mesenchymal stem cells from wharton's jelly after complete transection of the rat spinal cord." *PLoS ONE.* 3(10): e3336.

Zhang, W., Walboomers, X. F., Shi, S. et al. (2006). "Multilineage differentiation potential of stem cells derived from human dental pulp after cryopreservation." *Tissue Engineering.* 12(10): 2813-2823.

Zhang, Y.-N., Lie, P.-C. and Wei, X. (2009). "Differentiation of mesenchymal stromal cells derived from umbilical cord Wharton's jelly into hepatocyte-like cells." *Cytotherapy.* 11(5): 548-558.

Zou, Y., Misri, S., Shay, J. W. et al. (2009). "Altered States of Telomere Deprotection and the Two-Stage Mechanism of Replicative Aging." *Mol. Cell. Biol.* 29(9): 2390-2397.

Zuk, P. A., Zhu, M., Ashjian, P. et al. (2002). "Human adipose tissue is a source of multipotent stem cells." *Molecular Biology of the Cell.* 13(12): 4279-4295.

In: Advances in Genetics Research. Volume 12
Editor: Kevin V. Urbano
ISBN: 978-1-63321-351-7

Chapter 4

GENETIC DIVERSITY OF HUMAN AND ANIMAL PAPILLOMAVIRUSES

Antonio Carlos de Freitas[1*], Ana Pavla Almeida, Diniz Gurgel,[1] Bárbara Simas Chagas,[1] Valdir de Queiroz Balbino[2] and Marcus Vinicius Aragão Batista[2]

[1]Laboratory of Molecular Studies and Experimental Therapy, Department of Genetics, Center for Biological Sciences, Federal University of Pernambuco, Pernambuco, Brazil
[2]Laboratory of Bioinformatics and Evolutionary Biology, Department of Genetics, Center for Biological Sciences, Federal University of Pernambuco, Pernambuco, Brazil

ABSTRACT

The papillomaviruses form a highly diverse group that infects mammals, birds and reptiles. There is a huge veterinary and public health importance associated with these viruses. Because of this, human and bovine papillomaviruses are the most studied viruses of this group. Infection with human papillomavirus (HPV) represents the most important risk factor in the genesis of cervical cancer. Bovine papillomavirus (BPV) infects basal epithelial cells leading to the formation of benign and malignant tumors, when in synergism with action of environmental co-factors. Brazil is a country with continental dimensions; therefore the understanding of the genetic variability and the distribution of these viruses is essential for an active surveillance program and the design of more efficient vaccines and diagnostic methods. In what concerns the association of HPV and cervical cancer, HPV-16 is the most frequent type worldwide, with the exception of Indonesia where the HPV-18 happens to be the most common type. Although the HPV-16 is the most frequent, the second most important type differs between different regions in the world. So, this difference in the incidence of HPV genotypes needs to be elucidated, along with the genetic variability of these isolates that could lead to distinct

* Corresponding author: Antonio Carlos de Freitas, Department of Genetics, Federal University of Pernambuco, Cidade Universitária, CEP 50.670-901, Recife – Pernambuco – Brazil. E-mail: acf_ufpe@yahoo.com.br.

oncogenic potential among different variants and subtypes. Oncogenic variants of HPV-31 were found and some of the mutations were observed in regions involved in host immune recognition. Changes in amino acid sequences between variants from different geographical regions evidence the possibility of functional divergence between them, suggesting that the variants detected may have different adaptive fitness. On the other hand, twelve BPV types are known so far, but only three of them are associated with cancer. There is a great diversity of BPV types among cattle and high co-infection was found. The presence of infection with multiple viral types could lead to suppression of the immunological system making the clinical state of the animals worse. So, further molecular epidemiological investigations on the incidence and diversity of papillomavirus infection will assist in establishing a more accurate view of the distribution of these viruses. This could help the development of accurate and specific diagnostics, prophylactics and therapeutic requirements.

INTRODUCTION

Papillomaviridae Family

Papillomaviruses (PVs) are a large group of viruses that infects a wide variety of vertebrate species. Nowadays, it is widely known that PVs are involved with epithelial and mucosal lesions of several mammalian and non-mammalian groups, such as birds and reptiles [1]. However, the link between the infection caused by PVs and skin diseases was discovered only in the mid-of 20th century. The first article that reported the PVs and their association with papillomatosis was published in 1937, when the researcher Richard Shope observed that the Cottontail rabbit papillomavirus (CRPV, now SfPV1) suspensions might actively immunize rabbits against papillomatosis disease [2]. Furthermore, the researchers Peyton Rous and J. Beard published an article describing the transformation of papilloma lesion in squamous cell carcinoma in rabbits, thus concluding that papillomavirus is involved in skin cancer development [3]. Peyton Rous was awarded the Nobel Prize in Physiology or Medicine (1967) in recognition of his efforts to demonstrate the relationship between virus and cancer. In the early 1970s and 1980s, research groups published several articles demonstrating the involvement of Human papillomavirus (HPV) in cervical smear, genital warts and laryngeal papillomas [4]. However, only in 1983, zur Hausen and co-workers isolated HPV-16 in biopsies from cervical cancer [5]. To date, 151 HPVs have been described (http://pave.niaid.nih.gov/#home) and *classified according to the oncogenic potential in High Risk (HR) HPV and Low Risk (LR) HPV* [4]. Harald zur Hausen was recently awarded the Nobel Prize in Physiology or Medicine (2008) due to the discovery that HPV is "necessary but not sufficient to develop cervical cancer".

According to the World Health Organization (WHO), cervical cancer represents 13% of female cancer and it is the second leading cancer in women worldwide [6]. HPV is a major sexually transmitted infection (STI), since this viral group is involved in several cancers such as cervical, penis, vagina, vulva, anus, head-neck and non-melanoma skin cancers [7,8]. Furthermore, HPV infection is also involved in benign diseases, such as anogenital warts, laryngeal papillomas, and psoriasis [7,8]. In 2006, HPV vaccines against HPV-6, -11, -16 and -18 were approved. However, persistency and efficacy analysis, as well as selection of other HPV types, should be considered in future studies concerning PVs. STI is not exclusive to human and PV DNA was also detected in genital papillomas of cetacean and macaque species [9,10]. Thus, the medical importance of PVs infections and their relationship with skin diseases lead us to question how the molecular events of host-virus could contribute to diseases caused by this infection.

Papillomaviruses Characteristics

PVs are non-encapsuled and icosahedral virus group, measuring approximately 50-55 nm in diameter, and they contain a circular double-stranded DNA with approximately 8 kilobases (kb) in size. As mentioned previously, PVs infect the epithelial cells of several mammalian and non-mammalian species and can cause benign and malignant diseases [11–14]. Currently, PVs were detected in cats, dogs, cows, cattle, deer, macaques, turtles, snake, goat, rabbits and other animals (Figure 1).

Because their strong cell tropism, PVs are considered epithelialtropic and mucosotripic agents, although several studies have detected PVs DNA in non-epithelial sites, such as blood, semen (spermatozoa) and placent [15–22]. Also, PVs are normally considered host-specific, although some PVs can be found in several hosts [12–14]. For example, *Bovine papillomavirus* 1 (BPV-1) and BPV-2 can infect cow, cattle, buffalo and giraffe.

Figure 1. Some examples of PVs and their hosts, including mammalian, bird and reptile groups. *Human papillomavirus* (HPV) infects Humans; *Feline papillomavirus* (FdPV) infects domestic cats; *Bovine papillomavirus* (BPV) infects domestic cows and cattle; *Alces alces papillomavirus* 1 (AaPV1) infects European elk; *Capreoulus capreolus Papillomavirus* 1 (CcaPV1) infects Western roe deers; *Macaca fascicularis papillomavirus* (MfPV) infects Cynomolgus macaques; *Chelonia mydas papillomavirus* 1 (CmPV1) infects turtles; *Morelia spilota papillomavirus 1* (MsPV1) infects snakes; *Capra hircus papillomavirus* 1 (ChPV1) infects domestic goats; *Cottontail rabbit papillomavirus* (CRPV) infects rabbits.

Papillomaviruses Genome Diversity

PVs genome typically consists of 8 Open Reading Frames (ORFs), in which generally six ORFs are classified as early genes (E1, E2, E4, E5, E6 and E7) and two as late genes (L1 and L2), according to the expression of these genes in the viral development into keratinocytes cells. Early genes are expressed in the surface of epithelium and late genes in suprabasal layer epithelium. PVs genome are structurally conserved, although the E3, E4, E5/E5a and E5b, E8 and E9 genes may be present or absent in these genomes. Furthermore, PVs possess a non-coding region named Long Control Region (LCR), which is located in the upstream of the early and late genes (Figures 2, 3, 4 and 5).

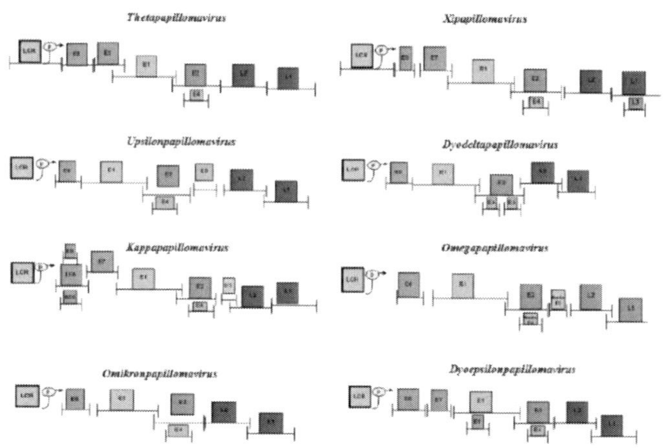

Figure 2. PVs genomes in a liner format of *Thetapapillomavirus, Xipapillomavirus, Upsilonpapillomavirus, Dyodeltapapillomavirus, Kappapapillomavirus, Omegapapillomavirus, Omikreonpapillomavirus and Dyoepsilonpapillomavirus genera*. Concerning the *Thetapapillomavirus* genus, do not was observed the both E6 and E7 genes. *Xipapillomavirus* genus do not have E6 gene. *Upsilonpapillomavirus, Dyodeltapapillomavirus, Omegapapillomavirus, Omikronpapillomavirus* genera do not have E7 gene in their genomes. The E9 gene are present in *Thetapapillomavirus* and *Dyoepsilonpapillomavirus* genera.

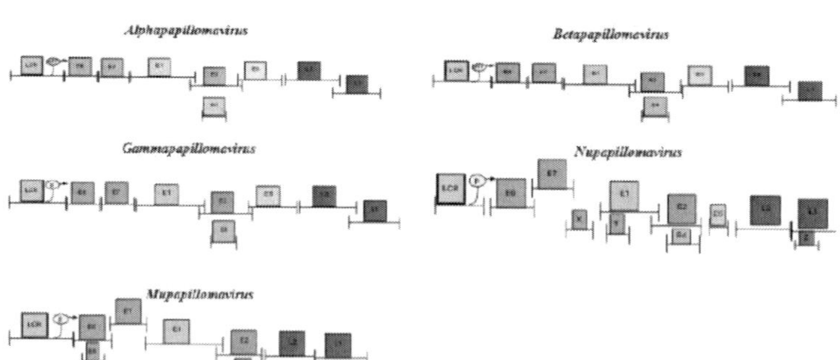

Figure 3. PVs genome in a liner format of *Alphapapillomavirus, Betapapillomavirus, Gammapapillomavirus, Mupapillomavirus and Nupapillomavirus* genera. These genera of virus infect the humans and monkeys and commonly presents E1, E2, E4, E5, E6, E7, L1 and L2 genes. However, the Nupapillomavirus present coding region called X, Y and Z, which possess na unknown function. The Mupapillomavirus possess E8 gene and do not possess the E5 gene.

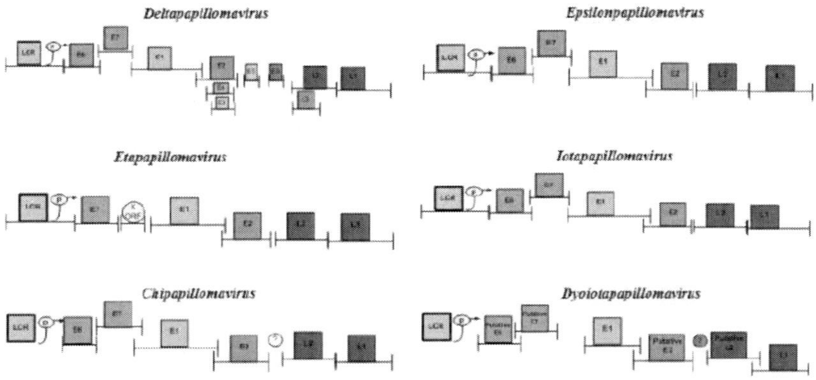

Figure 4. PVs genomes in a liner format of *Deltapapillomavirus, Epsilonpapillomavirus, Etapapillomavirus, Iotapapillomavirus, Chipapillomavirus, Dyoiotapapillomavirus* genera. PVs E3, E9 and L3 genes are present in *Deltapapillomavirus* genus and they do not still possess a known function. *Epsilonpapillomavirus, Etapapillomavirus, Iotapapillomavirus, Chipapillomavirus and Dyoitopapillomavirus* genera do not have E4 gene.

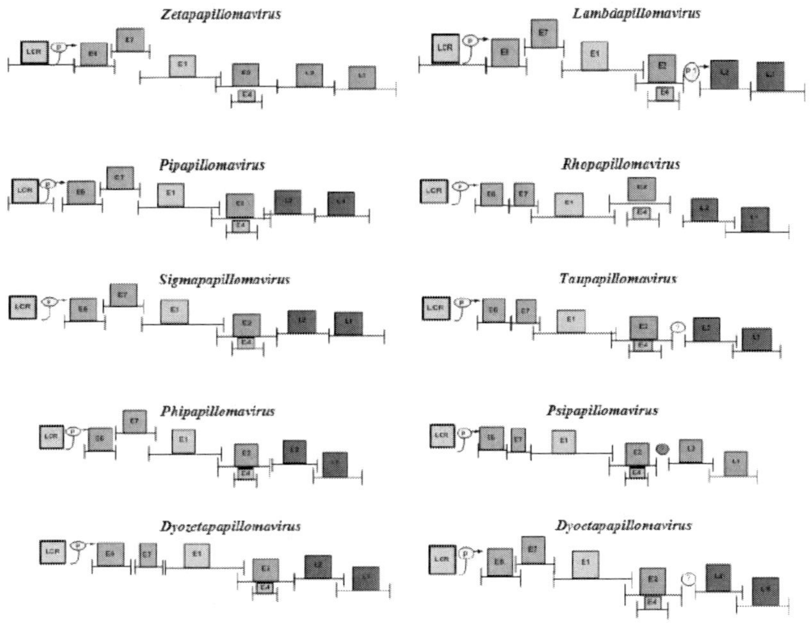

Figure 5. PVs genomes of *Zetapapillomavirus, Lambdapapillomavirus, Pipapillomavirus, Rhopapillomavirus, Sigmapapillomavirus, Taupapillomavirus, Phipapillomavirus, Psipapillomavirus, Dyozetapapillomavirus and Dyoetapapillomavirus* genera. These genera groups do not have E5 gene (Other genera do not present E5 gene, such as *Thetapapillomavirus, Omikronpapillomavirus, Dyoepsilonpapillomavirus, Mupapillomavirus, Epsilonpapillomavirus, Iotapapillomavirus, Etapapillomavirus, Chipapillomavirus and Dyoiotapapillomavirus*). The *Lambdapapillomavirus, Taupapillomavirus, Psipapillomavirus* and *Dyoetapapillomavirus* possess a non-coding region between E2 and L2 genes.

Classification of Papillomaviruses

Initially, Papillomaviruses and Polyomavirus were grouped into *Papovaviridae* family due to their small size, non-enveloped capsids and circular double-stranded genome features. However, during the 8th International Committee on Taxonomy of Viruses (ICTV), the division of the *Papovaviridae* in *Papillomaviridae* (constituting of PVs) and *Polyomaviridae* (constituting of Polyomavirus – PVD) families was determined due to their nucleotide sequence distance.

Although these viral groups have been divided, two papilloma/polyoma virus hybrids were detected, in which there are homologous sequences in early genes to polyomavirus and late genes to papillomavirus. However, these viral groups were not included in Papillomavirus [13].

Actually, the ICTV categorized Papillomavirus into family, genus, species, strain and type-species [13]. Furthermore, PVs were classified through the L1 nucleotide sequence identity, since there was not a recognition culture or serological methods to classify PVs yet. In 2010, *Papillomaviridae* family comprised 29 genera, 189 PVs complete genome deposited in GenBank, being 69 non-human PVs, three infecting birds and two infecting reptiles PVs [13]. To date, there are 151 HPV complete genomes deposited in GenBank http://www.ncbi.nlm.nih.gov/genbank/. PVs genera are written in Greek Alphabet, comprising the following: *Alphapapillomavirus*, *Betapapillomavirus*, *Gammapapillomavirus*, *Deltapapillomavirus*, *Epislonpapillomavirus*, *Zetapapillomavirus*, *Etapapillomavirus*, *Thetapapillomavirus*, *Lotapapillomavirus*, *Kappapapillomavirus*, *Lambdapapillomavirus*, *Mupapillomavirus*, *Nupapillomavirus*, *Xipapillomavirus*, *Pipapillomavirus*, *Rhopapillomavirus*, *Sigmapapillomavirus*, *Taupapillomavirus*, *Upsilonpapillomavirus*, *Phipapillomavirus*, *Chipapillomavirus*, *Psipapillomavirus*, *Omegapapillomavirus*, *Dyodeltapapillomavirus*, *Dyoepsilonpapillomavirus*, *Dyozetapapillomavirus*, *Dyoetapapillomavirus*, *Dyothetapapillomavirus*, *Dyoiotapapillomavirus* and *Omikronpapillomavirus* [13]. However, it is well known that the PV diversity is under-estimated and a large number of PVs will be detected in the next years. To avoid the confusion caused by the nomenclature, the ICTV still recommends the use of Greek Alphabet to new employed genera, by adding the prefix Dyo, such as *Dyodeltapapillomavirus*, *Dyoepsilonpapillomavirus*, *Dyozetapapillomavirus*, *Dyoetapapillomavirus*, *Dyothetapapillomavirus* and *Dyoiotapapillomavirus*.

Evolution of Papillomaviruses

The PV diversity can be explained by multiple evolutionary mechanisms [30]. Virus host-divergence is an important evolutionary force. However, this force solely cannot explain the evolution of PVs and their diversity. Thus, alternative mechanisms such as within-host virus duplication, recombination, viral sorting, or viral adaptation after a host switch, may contribute to explain the PV diversification [31–33].

Although it is not clear how these evolutionary mechanisms act on PVs diversity, some hypotheses were raised through the use of robust methods. Multiple evolutionary mechanisms such as co-evolution, recombination, adaptive radiation and interspecies transmission should drive the PVs diversification [30]. At least six PVs lineages should be associated with an ancestral mammalian host, thus giving rise to the PVs types that infect today's mammalian hosts [31]. The importance of alternative mechanisms other than co-divergence was suggested after the use of different statistical approaches to assess topological and branch-length congruence [32]. Batista et al. [34] used another entropy-based statistical approach to assess the evolution of PVs, showing that the genome has regions that could be used as markers to infer PVs phylogeny.

The understanding of PVs diversity is still limited, probably underestimated. Therefore, more new PVs types should be sequenced to increase our knowledge about PV evolution. The diversity of subtypes and variants could show a more detailed and refined scenario of PVs diversification, thus increasing insights into the representativeness of each PVs type [33].

Papillomaviruses Infection Cycle

The infectious cycle of PVs depends on the differentiation of keratinocytes [11]. The infection starts in the basal layers through micro-lesions or abrasions in the stratified epithelium, which occur in the skin or mucosa by direct contact [11]. From there, the infection could follow three ways: latent infection, clinical or subclinical manifestations [35]. After the entry of the virus into the cell, the viral genome is stabilized in the form of extra-chromosomal elements (episomes) in the nucleus and the number of copies is increased to approximately 50 to 100 per cell. When these infected cells divide, the viral DNA is equally distributed between the daughter cells. One of the infected cells migrates from the basal layer towards the upper layers of the epithelium in order to monitor the process of cellular differentiation of keratinocytes, while the others continue dividing in the basal layer, serving as a reservoir for viral DNA for subsequent cell divisions [36,37]. Although PVs infections occur in the basal layers, the production of the virus is restricted to the suprabasal layer cells, which are not lysed by the production of virions, thus continuing the proliferation. This dependent differentiation promotes the infection and the maintenance of PVs in the basal layers for periods up to several years [38].

From the time of the infection to the release of viral particles goes approximately three weeks, corresponding to the period required for differentiation and desquamation of the basal keratinocytes [39]. The period between initial infection and the appearance of papillomas may be related to the concentration and type of the PVs that infect the tissue [40]. These viruses can cause lesions on the surface of mucosal and cutaneous tissue. Typically, these lesions regress spontaneously although persistent infection of a subset of these viruses can lead to the formation of malignant tumors [41–44].

Due to the fact that HPV and BPV have a great medical and veterinary importance, this chapter will focus on the recent insights into HPV and BPV diversity.

Human Papillomavirus

Human papillomavirus (HPV) are a diverse group of virus that infect humans and monkeys and they are associated with several benign and malign diseases [13,41], notably cervical cancer. Cervical cancer is the second cause of gynecological cancer worldwide [45].

According to WHO, more than 500.000 new cases arise and 250.000 women die per year. Among the cases of cervical cancer, 85% are squamous cell cancer, followed by adenocarcinoma and small cell neuroendocrine tumor [7]. Actually, the prevention strategies to cervical cancer include papanicolaou analysis and excision of precancerous lesions. In the last years, the screening programs allowed the decrease in the number of cases worldwide, however cervical cancer remains as one of the major causes of death in female worldwide. In the developing countries, there is a high frequency of cervical cancer (86%) when compared to developed countries, probably due to inefficient screening programs [6].

Several studies have demonstrated that the HPV infections are associated with cervical cancer [6,41,46]. Also, HPV infection seems to be an important cofactor involved in other anogenital cancers. In 2010, statistical analysis of WHO showed that there are HPV infections in 75% of anal cancer patients [6]. Moreover, HPV-16 and HPV-18 were the most frequent type, detected in 73% of the anal cancer cases. Similarly, the HPV infection is involved in 40% of vulvar cancer, in which 36% of the patients are infected with HPV-16 and HPV-18 [6]. 66% of vaginal cancer patients are infected with HPV, notably by the HPV-16 and HPV-18. Penile cancer seems to have a relationship with HPV infection, since 46% of patients are infected with HPV and 35% could be infected with HPV-16 or HPV-18.

As mentioned previously, 151 HPV types were divided into five genera: *Alphapapillomavirus, Betapapillomavirus, Gammapapillomavirus, Mupapillomavirus* and *Nupapillomavirus* [http://pave.niaid.nih.gov/#home] [12]. In 2010, it was reported that the *Alphapapillomavirus* genus was constituted by 74 HPV types, *Betapapillomavirus* genus has 41 HPV types, *Gammapapillomavirus* has 16 HPV types, *Mupapillomavirus* has two HPV types and *Nupapillomavirus* has only one specie (see Table 2) [13]. *Alphapapillomavirus* and *Betapapillomavirus* can be found in humans and monkeys. *Aphapapillomavirus, Betapapillomavirus* and *Mupapillomavirus* are involved in benign and malign tumors, while the *Gammapapillomavirus* and *Nupapillomavirus* seem to be involved in benign diseases only [13]. All HPV types are distributed worldwide and can be transmitted by direct contact.

Among these genera, *40 HPV types infect* the *genital tract, being classified according to the oncogenic potential in High Risk (HR) HPV:* 16, 18, 31, 33, 35, 39, 45, 51, 52, 56, 58, 59, 68, 73, 82*; and Low Risk (LR) HPV:* 6, 11, 42, 44, 51, 53, 83 [4]. The most common HR-HPV types worldwide are the 16 (57%), 18 (16%), 58, 33, 45, 31, 52, 35, 59, 39, 51 and 56 [47]. About 90% of the LR-HPV infections are caused by the HPV-6 and HPV-11 (see Table 1).

HR-HPV infection is commonly transmitted by sexual intercourse and can cause lesions in epithelial tissues. These lesions can regress in 6-12 months [4]. However, if not treated, these lesions can progress from *pre-malign* to cervical cancer, such as: cervical intraepithelial neoplasia grade 1 (CIN1) (or mild dysplasia); intraepithelial neoplasia grade 2 (CIN2) (or moderate dysplasia); intraepithelial neoplasia grade 3 (CIN3) or *in situ* carcinoma; and squamous cell carcinoma or adenocarcinoma [4]. The Bethesda System is based on papanicolaou analysis and it classifies these pre-malign lesions into: atypical squamous cells of undetermined significance (ASCUS); and squamous intraepithelial lesion (SIL), which comprises the low-grade SIL (mild dysplasia) and high-grade SIL (moderate severe dysplasia and *in situ* carcinoma) [48]. The cervical intraepithelial neoplasia grade 2/3 regress in 28% of the patients in a short period of time [49].

Table 1. Characteristics of HPV from *Alphapapillomavirus*, *Betapapillomavirus*, *Gammapapillomavirus*, *Mupapillomavirus* and *Nupapillomavirus* genera. All HPV are found worldwide and can infect mucosa and epithelium through contact. *Gamapapillomavirus* and *Mupapillomavirus* HPV can cause begin diseases, such as warts and papillomas. On the other hand, *Alphapapillomavirus*, *Betapapillomavirus* and *Nupapillomavirus* can cause benign and malign diseases, notably cervical cancer, anal cancer, penile cancer and other

Genus	Geografical localization	Tissue tropism/Host	Transmission	Diseases associated	Types/Reference
Alphapapillomavirus	Worldwide	Epihtelium and mucosa Humans, Monkeys	Sexual contact, autoinoculation	Ginecological malignt and benign tumors	PpPV-1, HPV-2, HPV-3, HPV-6, HPV-7, HPV-10, HPV-11, HPV-13, HPV-14, HPV-16, HPV-18, HPV-28, HPV-31, HPV-26, HPV-27, HPV-29, HPV-30, HPV-31, HPV-32, HPV-33, HPV-34, HPV-35, HPV-39, HPV-40, HPV-42, HPV-43, HPV-44, HPV-45, HPV-51, HPV-52, HPV-53, HPV-54, HPV-56, HPV-57, HPV-58, HPV-59, HPV-60, HPV-62, HPV-66, HPV-67, HPV-68, HPV-69, HPV-70, HPV-71, HPV-72, HPV-73, HPV-74, HPV-77, HPV-78, HPV-81, HPV-83, HPV-84, HPV-86, HPV-87, HPV-89, HPV-90, HPV-82, HPV-85, HPV-91, HPV-94, HPV-97, HPV-102, HPV-106, HPV-117, MmPV-1, MfPV-3, MfPV-4, MfPV-5, MfPV-6, MfPV-7, MfPV-8, MfPV-9, MfPV-10, MfPV-11. [13]
Betapapillomavirus	Worldwide	Epihtelium and mucosa Humans Monkeys	Contact	Skin benign and malign tumors	HPV-5, HPV-8, HPV-9, HPV-12, HPV-14, HPV-15, HPV-17, HPV-19, HPV-20, HPV-21, HPV-22, HPV-23, HPV-24, HPV-25, HPV-36, HPV-37, HPV-38, HPV-47, HPV-49, HPV-75, HPV-76, HPV-80, HPV-92, HPV-93, HPV-96, HPV-98, HPV-99, HPV-104, HPV-105, HPV-107, HPV-100, HPV-110, HPV-111, HPV-113, HPV-115, HPV-118, HPV-120, HPV-122, HPV-124, CgPV-2, MfPV-1. [13]
Gammapapillomavirus	Worldwide	Epihtelium and mucosa Humans	Contact	Skin benign tumors	HPV-4, HPV-48, HPV-50, HPV-60, HPV-65, HPV-88, HPV-95, HPV-101, HPV-103, HPV-108, HPV-109, HPV-112, HPV-116, HPV-119, HPV- 121, HPV-123. [13]
Mupapillomavirus	Worldwide	Epihtelium and mucosa Humans	Contact	Skin benign tumors	HPV-1, HPV-63 [13]
Nupapillomavirus	Worldwide	Epihtelium and mucosa Humans	Contact	Skin benign and malign disease	HPV-41[13]

HPV Genome

Similarly to the PVs, HPV DNA has eight open reading frame (ORFs), namely E1, E2, E4, E5, E6, E7 (expressed in early phase of infection), L1, L2 (expressed in late phase of infection) and a control region denominated long control region (LCR) [4,11] (Figure 3). E1 and E2 ORFs are involved in viral DNA replication. The E4 gene is implicated in maturation of HPV virion. The E5, E6 and E7 genes are involved with transformation activities, in which E6 and E7 are involved with integration of HPV *DNA* within the cell DNA. Thus, HPV DNA can immortalize human keratinocytes due to the E6 and E7 oncogene interactions with p53 and pRb suppressor tumor proteins, respectively. These interactions inhibit the apoptosis process (see above).

L1 and L2 are involved in capsid formation. Moreover, HPV DNA possesses a LCR, which regulates the transcription of the E6 and E7 viral oncogenes through transcription factors of the virus and the host cells [11] (Figure 6).

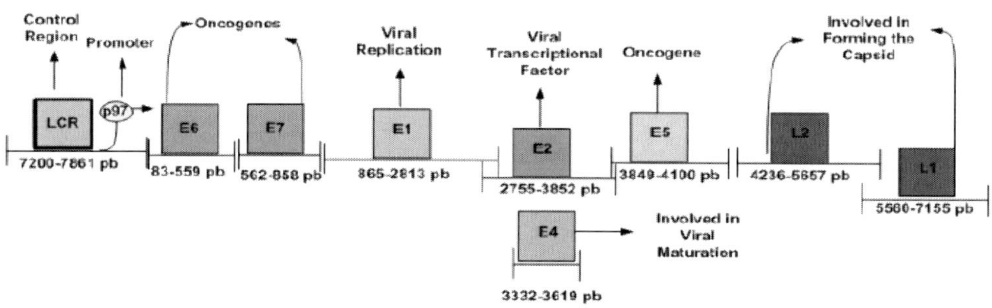

Figure 6. HPV-16 complete genome is shown in a linear format (based in RefSeq NC_001526.2). After the break in stratified epithelium tissue, the HPV can infect basal cell layer, thus starting an infection process. The *entry of HPV* into the *host cell* is initiated by the binding of the virus particles into the cell surface receptors, such as the heparan sulfate (not shown). Subsequently, there are expression of E1, E2 (to replication of viral DNA), E4 (probably involved in maturation of virion), E5, E6 and E7 genes (involved with transformation activity and immortalization), which results in replication of the HPV DNA as an *episome*. L1 and L2 genes are expressed in a late stage, thus resulting in viral capsid formation. Finally, the HPV virion is released at the cell surface, which can result in infection of other sites.

Integration of Human Papillomavirus in Host Genome

Sexually active women are at risk for genital infection by HPV. Most genital HPV infections regress and only a minority of women will develop persistent HPV infection that could cause cervical intraepithelial neoplasia (CIN) which is a precursor of invasive cervical cancer. The integration of high oncogenic risk HPV genome into the host chromosome is considered to be a key event in the progression of CIN to invasive cervical cancer [11], occurring downstream of the early genes E6 and E7, often in the E1 or E2 region. This disruption results in the loss of negative feedback control of oncogene expression by the viral regulatory E2 protein [10].

This event is often followed by over-stimulation of the transcription of genes E6 and E7 that disrupt the cell cycle through interaction with different proteins, such as the product of the tumor suppressor gene (p53) and the retinoblastoma protein (pRb) inducing viral immortalization capacity, and a high level of chromosomal instability (Figure 7) [50,51].

Several studies of HPV DNA integration have focused mainly on HPV-16 since it is considered the type that confers the highest risk of cervical cancer and also the type most commonly detected in women with normal cervical cytology [52]. In low-grade cervical lesions, the majority of HPV genomes persist in episomal state, whereas in high-grade lesions and invasive carcinomas have integrated HPV DNA [53].

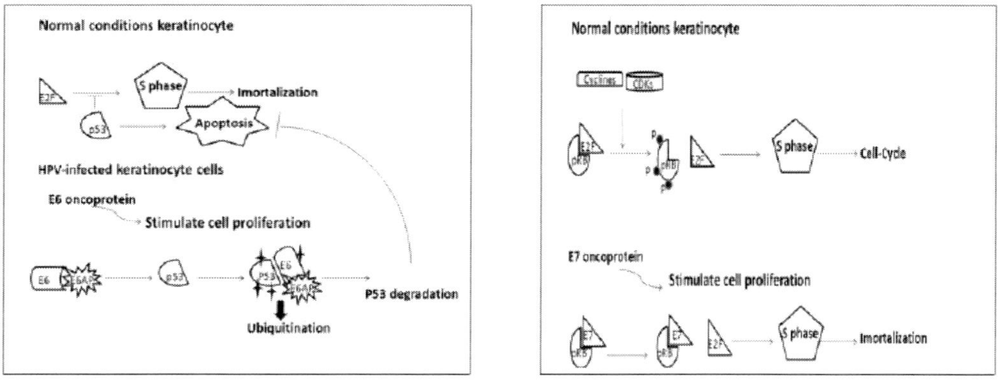

Figure 7. Transformation and immortalization effects of HPV E6 and E7 oncoproteins in keratinocyte cells. Briefly, HPV-E6-AP complex is able to *target p53* for *degradation* via the *ubiquitin-proteasome* system, leading to indeterminate division and immortalization of cells. Furthermore, E7 oncoprotein also leads to immortalization of the cell due to the regulation of pRB, releasing the E2F activity.

HPV Distribution

HPV-16, HPV-18, HPV-31 and HPV-52 are the four most frequent types found in health women (without lesions) worldwide [6]. However, the frequency of HPV types change according to grade lesion cases, cervical cancer cases, and the where the patients live. For example, in low grade lesion, the fourth most frequent HPV types are 16, 51, 31 and 52, while in high grade lesion the fourth most frequent HPV types are 16, 31, 58 and 33 [6]. On the other hand, the HPV-16, 18, 58 and 33 are the most frequent types in patients with cervical cancer worldwide [6]. Furthermore, the HPV types associated to cervical cancer in developing countries and developed countries are different. In the developing countries, the most frequent types are the HPV-16, 18, 58 and 33, while in developed region the most frequent types are the HPV-16, 18, 33, and 31 [6]. The contrast in the prevalence between the HPV types seems also to occur concerning the cancer type. Indeed, the most common HPV types in squamous cell cancer are the species 16, 18, 58 and 33. Moreover, in adenocarcinoma cancer, the most frequent HPV types are 18, 16, 45 and 33 [6].

According to WHO, the prevalence of HPV in health women in Latin American and Caribbean is 17%. In addition, the prevalence of HPV in women with low grade, high grade lesion and cervical cancer are 22%, 46% and 66%, respectively [6]. In Brazil, the prevalence of HPV in women with normal cytology is 14%. Concerning the cancer type, the four most prevalent HPV types in squamous cell carcinoma in Brazil are HPV-16, 18, 35 and 45 [54]. On the other hand, HPV-18, 16, 33 and 35 types are the most frequent in adenocarcinoma

[54]. Also in Brazil, HPV-16 is the most prevalent virus involved in carcinogenic activity; however, the second most important type changes according to the Brazilian regions. For example, in Northeastern Brazil the four most commonly HPV types are 16, 31, 33 and 18 [55]. In contrast, in Midwest region, the four most commonly HPV types are 16, 18, 31 and 33 [56]. Thus, differences on the prevalence of HPV types might have negative impact in adoption of public policy, concerning mass vaccination.

HPV Genetic Variability and Evolution

An HPV type differs from another in at least 10% nucleotide sequence of the L1 gene. The additional term "subtype" is used to identify HPV genomes with L1 nucleotide sequences that differ between 2-10% of that of the closest type, and the varying types differ in less than 2% in nucleotide sequence of L1, and 5% in LCR [12].

Substantial data demonstrate the existence of a natural diversity of variants of a determined type of HPV [56,58]. Different HPV types have distinct oncogenic potential and, the same way, intratypic variants may also have differential oncogenicity causing the development of cervical cancer [60–62]. Moreover, these differences may contribute to disparities in the incidence of cervical cancer worldwide [63].

Studies related to genetic variability revealed that different variants of HPV-16 and HPV-18 co-evolved with the three major human phylogenetic branches: African, Caucasian and Asian. However, HPV-16 variants were grouped into five distinct groups distributed in different geographical regions, such as European (E), Asian (As), Asian-American (AA), an African 1 (Af-1) and African 2 (Af-2) [64–66]. The variants of HPV-18 were properly grouped into three distinct groups: European (E), Asian-American (AA) and African (Af) [67–69]. In relation to HPV-31, it is suggested that HPV-31 variants do not exhibit the same patterns above mentioned with respect to the co-evolution with human ethnic groups [58,70]. However, Northeastern Brazil is characterized by strong ethnical admixture, possibly explaining isolates distribution in different clades of the phylogenetic trees. Another explanation would be that these HPV-31 variants might have existed before the spread of humans and the clusters specific for the ethnic groups arose subsequently, as suggested by Calleja-Macias et al. [63,70].

The possible variations in the nucleotide sequences of HPV have been used as an important tool to assist in epidemiological studies of transmission and viral persistence [71]. The differences in nucleotide sequences between the variants of one type of HPV can result in changes in amino acids encoded and can lead to distinct oncogenic potential between the different variants. Variations in the nucleotide sequence of the LCR region, which contains sequences important for the regulation of viral replication and transcription of early genes could result in loss of the binding capacity of a specific transcriptional factor, as well as changing its recognition site and the affinity of a specific transcription factor [72]. With respect to the L1 region, one or more amino acid changes could lead to a conformational change in the capsid protein, interfering with the conformation of epitopes relevant for viral neutralization and interference in vaccine strategies [73]. It can also be said that the importance attached to nucleotide variations in E6 and E7 genes is directly related to the occurrence of variations in critical regions for the degradation of p53 and pRb, respectively, and the regions involved in immune recognition of the host [74].

Due to the high prevalence of HPV-16 and HPV-18 in the world, not only among asymptomatic women, but also in samples of cervical cancer, several epidemiological studies have been conducted to describe the molecular variants of HPV types of high oncogenic risk [75–83]. Studies suggest that European variants are associated with a lower risk of developing high-grade lesions, whereas non-European variants are associated with an increased risk of incidence and prevalence of these lesions [79,82,84–86]. Moreover, it has been demonstrated an association between non-European variants, both HPV-16 and HPV-18, and an increased risk of persistent infections [73,85].

Genetic variability studies performed on HPV-16 suggested that each viral variant is associated with risk of cervical neoplasia. Londesborough et al. [78] studying the genetic variability in HPV-16 E6 gene in women from UK reported a nucleotide substitution of thymine for guanine at the position 350 of the gene, leading to a substitution of valine for leucine at position 83 regarding the amino acid sequence of this protein. It was suggested that there was an association between this change and an increased risk of persistent infection and cytological progression to cervical intraepithelial neoplasia of grade 2/3 [78]. In studies performed by Kammer et al. [87] and Zehbe et al. [83], this type of change in E6 was associated with an increased risk of cervical disease progression compared with the prototype strain. de Boer et al. [88] found a high frequency of nucleotide changes in the virus type 16 in cervical cancer patients from Indonesia. Among those changes stands the nucleotide variation adenine for guanine at the position 276 of E6 ORF, which promotes the change of the amino acid asparagine to serine at the position 58 of E6 protein [89,90], describing the same change in cervical intraepithelial neoplasia and cervical cancer. Another study reported a change in the nucleotide position 647 in the E7 gene of HPV-16, with a change in the amino acid asparagine to serine at the position 29, and this variation in E7 was more frequent in samples of cervical cancer compared with precursor lesions, suggesting alterations in the biological properties of this variant [91].

With respect to genetic variability of E6 of HPV-18, studies have shown the nucleotide change C491A, which revealed the amino acid change N129K. This codon 129 is highly conserved within genotypes oncogenic HPV [56,92]. However, in vitro analysis has shown that this change to E6 does not interfere with their ability in promoting p53 degradation [93]. Variations in LCR of HPV-18 were also observed, among these variations A41G and T104C [56]. Both variations in this region are associated with high activity of the E6/E7 promoter by modulating Sp1 and YY1 activities [94]. The substitution observed in the codon 149 of L1 gene in HPV-18, is located near a linear epitope, which is present in the surface of the capsid and in common among many HPV genotypes [56,95]. Changes in the L1 gene may compromise the fitness of the virus and may be important in discriminating the infectious potential of different variations and in defining epitopes relevant to vaccine design [73,96,97]. Studies report that specific variants of HPV-18 may increase the oncogenic potential, which may represent an additional increase in the number of patients with cervical cancer in different regions [69,84,86,98,99].

Considering the genetic variations of HPV-31 E6, Chagas et al. [70] detected nucleotide changes at positions 213 and 413, which were previously described [58,101], however changes at positions 141, 190 and 368 were described for the first time in his study, where changes at positions 190, 368 and 413 caused amino acid exchanges in codons 64 (T to A), 123 (K to R) and 138 (A to V), respectively. Still in this work, all mutations found in E7 gene of HPV-31 (positions 67, 111, 136 and 184) were described for the first time, among which

67, 136 and 184 were classified as non-synonymous, causing amino acid changes in codons 23 (H to Y) 46 (E to Y) and 62 (K to E), respectively. Many of observed genetic variations in E6 and E7 HPV-31 were located in T-cell and B-cell epitope sites [70]. These variations may influence the presentation of viral peptides to the T-cell, being this an important mechanism for the control of infection and development of cervical lesions. The recognition of T-cell determinants by T cells (particularly T helper cells) contributes significantly to the cell-mediated immune response against infectious organisms, as T cells cooperate with B cells in the induction and maintenance of an effective antibody response and contribute to the maturation of cytotoxic T cells by interacting with macrophages [102]. Nucleotide and amino acid analysis were also performed to L1 gene of HPV-31, where it was revealed the presence of variants nucleotide. Of these, nine led to amino acid changes in L1 protein. The mutations identified with higher prevalence were the two synonymous A6943C and T6949A. This last mutation was associated with high grade cervical lesions [103]. Analysis of HPV-31 LCR sequence variations performed by Cento et al. [103] revealed some substitutions (G7449A, G7457A, C7474T, G7525A and T7575C) which potentially affected biding sites for transcription factors. In particular, the nucleotide substitution (G to A) at position 7749 may cause a gain of YY1 and a loss of SRY binding sites. An SRY binding site may be also affected by the T7659G mutation. Similarly, the mutation A7450C may cause the loss of the TATA binding site. Other variations at nucleotide positions 7457 (G to A) and 7474 (C to T) may result in a gain of YY1 biding site and in a loss of E2 biding site, respectively [103].

One early PV lineage, probably with mucosal tropism, originated the supertaxon comprising by *Alphapapillomavirus* and *Omicronpapillomavirus*, which includes viruses infecting primates, carnivores, swine and cetaceans. *Alphapapillomavirus* have viruses associated with cervical cancer in humans [105]. A second early PV lineage gave rise to a clade of cutaneotropic PVs including *Betapapillomavirus* and *Xi-papillomavirus*, that comprise primates, rodents, carnivores, ruminants, hedgehogs and marsupials [14]. Many viruses of this supertaxon have found in healthy skin and are apparently not associated with lesions [106]. Others cause malignant transformation depending on the genetic susceptibility of the host, for example, infections by HPV-5 or HPV-8 are associated with the development of squamous cell carcinoma in individuals suffering from the hereditary disease epidermodysplasia verruciformis [107]. A fourth PV lineage gave rise to the supertaxon including *Kappapapillomavirus* and *Lambdapapillomavirus*, with primates, rodents, lagomorphs and carnivores but not humans as hosts [14].

Bovine Papillomavirus

Bovine papillomaviruses (BPVs) are a heterogeneous group of viruses that infect bovine cattle and other close related animals. They cause bovine papillomatosis, a worldwide distributed veterinary disease, which leads to important economic losses. The world cattle population has exceeded 1 billion head of cattle. India, Brazil and China have the largest cattle population in the world, whereas USA, Brazil and China are the largest beef producer, respectively [108]. India has around 320 million head of cattle and Brazil has about 200 million head of cattle, however the majority of Indian cattle population is non-commercial. In 2011, it was produced nearly 57 million tons of bovine meat. USA produced about 12 million tons of bovine meat and Brazil produced around 9 million tons of bovine meat [108]. The world's cow milk production is around 583 million tons in 2009. Brazil is the fifth milk producer in the world with about 29 million tons in 2009 [109].

Bovine papillomatosis leads to economic depreciation of animals, deterioration of the animal leather, decrease of milk production through mastitis, and decrease of beef production through weight loss and death by cancer. The lesions may progress to cancer due to the synergistic action of genetic or environmental co-factors [110, 111].

Currently, 12 BPVs are known and classified into three genera: *Deltapapillomavirus* (BPV-1 and -2), *Epsilonpapillomavirus* (BPV-5 and -8) and *Xipapillomavirus* (BPV-3, -4, -6, -9, -10, -11 and -12). One BPV type is still unclassified (BPV-7) [12, 13, 111]. All BPV types characterized so far are associated with different histopathological lesions. *Xipapillomaviruses* are pure epitheliotropic (causing true papillomas), *Deltapapillomaviruses* induce fibropapillomas and *Epsilonpapillomaviruses* appear to cause both fibropapillomas and true papillomas [110]. Members of *Deltapapillomavirus* genus are also associated with sarcoids in horses and other animals [112, 117]. As far as we know, three types of BPV (-1, -2 and -4) are associated with cancer. When connected with environmental factors (e.g. bracken fern ingestion), BPV-1 and -2 cause urinary bladder tumors, and BPV-4 causes upper alimentary tract cancer.

The genome of all PVs are very similar, however some interesting features are noted in BPV genomes. First, unlike HPV, the major transforming protein in BPVs is E5. In *Xipapillomavirus* genus, E6 gene is replaced by an E5-like gene (initially defined as E8). Nevertheless, the knowledge that most functions of this protein are shared with the E5 protein of BPV-1 prompted its redefinition as E5 protein, not E8 [44, 119]. Similarly to most PVs types, BPV types that belong to *Deltapapillomavirus* and *Epsilonpapillomavirus* still retain the E6 gene [110]. In addition, other unusual genes are present in some BPV types: E3 (a small putative gene) in BPV-2 and -4; L3 (a small ORF that overlaps with 3' half of L1 or L2 gene) in BPV-3, -4 and -12; and L4 (a small putative gene) in BPV-4. The significance of these genes is still unclear.

Epidemiology

Papillomas are more often found in cattle than in any other domestic animal. In Brazil, for instance, there is a high prevalence of this disease in cattle in all regions of the country. Bovine of any age could be affected, however calves and yearlings have higher incidence, 46.21% in one study [122]. It is likely that specific factors like nutrition, ticks infestation and age contribute to this incidence. Also, the immune state of the affected cattle population influences the pathology. Among the breeds, Holsteins present a higher occurrence of the disease, although Olson et al. [123] observed that there was not a strong genetic influence in experimentally induced papillomas. Papillomatosis in dairy cattle seems to prefer udder and teats areas in adult animals and heifers. In calves, the most affected sites are: the neck, dewlap, muffle, ears, muzzle, periocular and perilabial regions [124–126].

The transmission between animals is probably made by fomites: contaminated milking equipment, halters, nose leads, grooming and earmarking equipment, rubbing posts and wire fences, and other objects contaminated by contact with affected cattle. Warts are rare in animals that are artificially inseminated, so sexual transmission of these lesions is probable in cattle. The way the cattle are held influences the incidence of bovine papillomatosis. Housed cattle are more often affected with bovine papillomatosis than cattle on pasture. Some BPV hosts could get naturally infected with BPV after exposure to diseased cattle or contaminated fomites in [125, 122]. Ticks are probably a major problem in bovine papillomatosis transmission. Willian et al. [1992] observed that all the diseased animals were infested with ticks. A study showed that some tick control measures reduced the incidence of bovine papillomatosis [55, 127].

Bovine papillomavirus is widespread in normal cattle skin and dermal tissues. The epidemiological importance of these latent infections is to be elucidated, however warts may grow, at sites of injury, through the reactivation of the latent DNA by physical irritation, immune depression and suppression. The lesion formation may occur in the absence of any previous clinical symptom, indicating a latent infection [122, 128].

Until recently, it was believed that BPVs were associated with distinct lesions, in different anatomic sites. BPV-1, -2 and -5 infect both mesenchymal and epithelial cells, causing common cutaneous warts and fibropapillomas [122]. On the other hand, BPV types that are members of *Xipapillomavirus* genus induce non-fibroblast epithelial and cutaneous lesions [122]. According to Borzacchiello and Roperto [110], BPV-1 causes teats and penile fibropapillomas, and urinary bladder tumours; BPV-2 is associated to cutaneous warts (usually on the forehead, neck, upper thorax and back), alimentary fibropapillomas and urinary bladder tumours; BPV-3 causes cutaneous papillomas [129]; BPV-4 is associated with pure epithelial papillomas of the upper gastrointestinal tract; BPV-5 induces rice grain fibropapillomas of the udder; BPV-6, frond papillomas of the teats; BPV-8 causes cutaneous papillomas; BPV-9 and -10 are associated to epithelial squamous papillomas of the udder.

In addition, BPV-4 is a viral type related to mucosal infections and it is responsible for upper digestive tract cancer in cattle when associated with bracken fern (*Pteridium aquilinum*). However, BPV-4 was already detected in skin lesions [52, 130], BPV-5 and -6 are usually associated with skin lesions in udder and teats, but they were found in lesions of the shoulder clod and around the eye of some animals, unusual sites for these viral types [130].

Although BPVs are considered an exclusively infecting epithelial cells group of viruses, there is evidence for the presence of BPV DNA, RNA and proteins in peripheral blood mononuclear cells [110, 131–137], placenta and amniotic liquid [132, 136], gametes [131,138,139], urine [139], uterus [133,136] ovarium [133], semen [133, 139, 140], also in milk and colostrum [132, 139]. These findings could support the hypothesis of the BPV vertical transmission [130–132, 141].

BPV Types Distribution

Bovine papillomavirus has been detected in cattle worldwide. Several studies showed the incidence of BPV in all continents [33]. BPV-1 is a viral type frequently found in skin lesions [130,131,132]. In Germany, BPV-1, -6, -8 and -10 were found in bovine warts cowshed [142]. Heifers presented benign teat tumors caused by BPV-6 in Japan [143]. BPV-1, -3, -5, -6, -8, -9 and -10 were detected in papilloma specimens, also in Japan [113,144,145]. In India, bovine cutaneous warts caused by BPV-1 and -2 were reported [119,146]. Recently, [147] identified BPV-10 in teat warts from cattle at a dairy farm also in India. BPV was found in equine sarcoid in western USA and Canada [148, 149]. In Africa, BPV was found in skin lesions of zebra, giraffe and antelope [118, 120]. In addition, BPV-4 and -2 were detected in cattle in Australia [52, 150].

Cattle from Brazil have also been investigated for the presence of BPV. BPV-1, -2, -6 and -8 were identified in skin warts of cattle from southern Brazil [151–153]. In northeastern Brazil, ten different types of BPV were also detected in skin lesions. Despite BPV-3 is rarely reported in the world, since the majority of studies about BPV infection focus on BPV-1 and -2, there is a high incidence of BPV-2 and -3 in northeastern Brazil [130]. BPV was extensively detected in other tissues in Brazil [139, 140, 154, 155].

BPV is also associated with cancer in cattle. BPV-4 infection and associated tumors of upper alimentary tract have been found in Brazil, the Nasampolai Valley of Kenya, Western Highlands of Scotland and in southern Italy [156, 157]. BPV-1 and -2 associated field cases of urinary bladder cancer in cattle were reported in continental Europe, Azores Islands, and some regions of Kenya, Brazil, New Zealand, India and China [110].

Some reports describe the occurrence of co-infection with different types of BPV worldwide [119, 130, 136, 139, 142, 144, 158, 159]. In Brazil, a high co-infection rate was found. The animals were constantly affected by diverse warts over the body, which could be a consequence of co-infection. This infection with multiple BPV types could prevent the regression of the lesions. Besides that, the presence of infection with multiple viral types could lead to the suppression of the immunological system making the clinical state of the animals worse [130].

Genetic Diversity and Evolution

The PVs is a very diverse group of viruses, which comprises more than 150 genomes of HPV types, more than 80 genomes of non-human Papillomavirus types, but less than 50 host species have at least one PVs associated. Inside this context, only 12 BPV types are known so far, which is a limited number when compared to HPV types. However, 14 new putative types were isolated and probably will increase considerably that number [144, 153, 160]. That shows the great genetic diversity of these viruses, as PVs types are defined based on L1 gene sequence dissimilarity [12].

Besides that, PVs are also classified into subtypes and variants, which are important to show a more detailed and refined scenario of PV diversification, increasing insights into the representativeness of each PVs type [33]. The differences found in nucleotide sequences of variants of PVs could be responsible for changes in the oncogenic potential, cellular location, host immune response, function of PVs proteins (affecting pathogenesis), or binding capacity of a transcription factor [70, 72, 73, 161]. The role of the nucleotide polymorphisms in BPV genomes is still scarce. However, they clearly have a great biological and evolutionary importance, based on HPV studies.

A few studies related to BPV genetic variability were done. The majority of these studies were associated with equine sarcoid, aiming at understanding differences between the disease in bovine and equine [154, 157]. A novel BPV-11 subtype was identified in Brazil, but the role of this genetic variation is still unknown [130]. Seven different BPV-1 E5 variants were detected in equine sarcoids, one of them was present in all isolates from Europe, a possible synonymous polymorphism characteristic of European variants of BPV-1. One polymorphism was found to alter an aliphatic side chain amino acid residue, suggesting that it could have important implications for the structure of E5 protein. An interesting observation is that most of the sequence variation found in this study was associated with an adaptation to the codon usage in mammalian cells [163].

Five different BPV-1 LCR variants and six different BPV-1 E2 variants were detected in Europe. Three non-synonymous sequence changes occurred within the transactivation domain of E2 and two within the 'hinge' region. In this study, specific BPV-1 LCR and E2 variants were infecting horses, suggesting that sarcoids are linked to variant BPV-1 genomes that preferentially infect horses and are kept within equids [164]. In another study, putative BPV-1 variants from Poland were divided into three phylogenetic groups and one separate isolate, showing the great genetic diversity of this virus in equine sarcoids [165]. All those findings suggest that the genetic variability found in BPVs could lead to different pathological processes, as well as it could increase the adaptation to other hosts. So, it is very important to study and understand the evolutionary mechanisms that guide the diversification of PVs.

In the evolutionary point of view, BPV are found in at least three distantly related lineages. First, BPV that belong to *Deltapapillomavirus* and *Epsilonpapillomavirus* form a paraphyletic group with some ovine PVs (OvPV-1 and OvPV-2), which infects a close related host. Other PVs that infect Artiodactyla are also close relatives of those BPVs. However, this group is clustered together with some equine and canine PVs. Second, BPVs that belong to *Xipapillomavirus* are grouped together with caprine PVs (ChPV-1). This group is related with a large cluster that involves some human, canine and rodent PVs. So far, BPV-7, an unclassified PV, has an uncertain phylogenetic position, which makes it difficult to infer its relatedness to other PVs. Therefore, all this diversity found in PVs that infect one host (*Bos taurus*) is a case of evolutionary incongruence between host and PV phylogeny, indicating that co-divergence alone cannot explain the PV diversity [30–32, 34, 166].

Some conflicting phylogenetic positions within *Xipapillomavirus*, which include BPV-3, -4, -6, -9, -10, -11 and -12, have been shown when analyzing early or late genes phylogenies [166, 167]. In general, the topological inconsistency between early and late genes phylogenies have been explained with ancient recombination events [30, 31, 166]. This also could be the explanation for the contradicting positions of BPV within *Xipapillomavirus*. For BPV, at least three lineages seem to originate the currently known types. These lineages probably passed through a prior divergence process preceding the host divergence. This could also explain the evolutionary proximity of BPV to PVs that infect distantly related hosts [33]. In addition, zoonotic transmission of PVs is a rare event but it occurs in BPV as they were found in zebras, horses, buffaloes and other animals [117, 118, 166, 168]. We could speculate that such a host switch could lower the evolutionary pressures on BPVs in cattle and may allow changes in infection timing and virulence in other hosts [169]. Therefore, other evolutionary mechanisms could be associated with BPV diversification. However sampling is still a limiting factor [33].

CONCLUSION

PVs are a diverse viral group that can infect humans, mammalian non-human, birds and reptile species. Nowadays, there are more than 200 PVs identified, which are divided into 29 genera. Basically, the genome of PVs presents genes that are expressed in an early stage and late stage. Furthermore, the PVs can present more than one non-coding sequence, named long control region (LCR). All the PVs present E1, E2, L1 and L2 genes. On the other hand, E3, E4, E5, E6, E7, E8 and E9 can be present or absent, according the genus.

To date, there are 151 HPVs and 12 BPV types described. Due to their medical and veterinary importance, HPV and BPV are the most studied viral groups. Concern to HPV, this viral group is the major cofactor involved in anogenital and non-anogenital cancers. HPV-16 is most prevalent specie in cervical, anal, vulvar, vagine and penile cancers worldwide. However, the subsequent most prevalent HPV types depend on several conditions, such as grade cervical lesion, cervical cancer types and the region where the patients live. Furthermore, variation in nucleotide sequence of HPV can result in changes in amino acid sequences, which can be the important to their oncogenic potential. Thus, these differences in prevalence of HPV types and intratype variability can have an important impact in adoption of public policies. Similarly, BPVs are involved with malign tumors in bovines together with environmental cofactors. As mentioned in HPV, changes in nucleotide sequence of BPVs can result in changes in amino acid sequences, which can lead to oncogenic potential differences. Thus, the understanding of the genetic differences among PVs, notably HPV and BPV, could explain why some HPV or BPV types are most oncogenic while other do not.

REFERENCES

[1] Lange CE, Favrot C, Ackermann M, Gull J, Vetsch E, Tobler K. Novel snake papillomavirus does not cluster with other non-mammalian papillomaviruses. *Virology Journal* 2011;8:436.

[2] Shope RE. Immunization of rabbits to infectious papillomatosis. *J Exp Med* 1937;65(2):219-231.

[3] Rous P, Beard JW. The progression to carcinoma of virus-induced rabbit papillomas (shope). *J Exp Med* 1935;62(4):523-548.

[4] zur Hausen H. Papillomaviruses and cancer: from basic studies to clinical application. *Nat. Rev. Cancer* 2002;2(5):342-350.

[5] Dürst M, Gissmann L, Ikenberg H, zur Hausen H. A papillomavirus dna from a cervical carcinoma and its prevalence in cancer biopsy samples from different geographic regions. Proceedings of the National Academy of Sciences 1983;80(12):3812 -3815.

[6] Who | strengthening cervical cancer prevention and control [s.d.].

[7] Hoory T, Monie A, Gravitt P, Wu T-C. Molecular epidemiology of human papillomavirus. *Journal of the Formosan Medical Association* 2008;107(3):198-217.

[8] Pfister H. Chapter 8: human papillomavirus and skin cancer. *J. Natl. Cancer Inst. Monographs* 2003;(31):52-56.

[9] Gottschling M, Bravo IG, Schulz E, Bracho MA, Deaville R, Jepson PD, Van Bressem M-F, Stockfleth E, Nindl I. Modular organizations of novel cetacean papillomaviruses. *Mol. Phylogenet. Evol.* 2011;59(1):34-42.

[10] Wood CE, Chen Z, Cline JM, Miller BE, Burk RD. Characterization and experimental transmission of an oncogenic papillomavirus in female macaques. *J Virol* 2007;81(12):6339-6345.

[11] Doorbar J. The papillomavirus life cycle. Journal of Clinical Virology 2005;32:7-15.

[12] de Villiers E-M, Fauquet C, Broker TR, Bernard H-U, zur Hausen H. Classification of papillomaviruses. *Virology* 2004;324(1):17-27.

[13] Bernard H-U, Burk RD, Chen Z, van Doorslaer K, Hausen H zur, de Villiers E-M. Classification of papillomaviruses (pvs) based on 189 pv types and proposal of taxonomic amendments. *Virology* 2010;401(1):70-79.

[14] Bravo IG, de Sanjosé S, Gottschling M. The clinical importance of understanding the evolution of papillomaviruses. *Trends in Microbiology* 2010;18:432-438.

[15] Bodaghi S, Wood LV, Roby G, Ryder C, Steinberg SM, Zheng Z-M. Could human papillomaviruses be spread through blood? *J Clin Microbiol* 2005;43(11):5428-5434.

[16] Kay P, Allan B, Denny L, Hoffman M, Williamson A-L. Detection of hpv 16 and hpv 18 dna in the blood of patients with cervical cancer. *J. Med. Virol.* 2005;75(3):435-439.

[17] Pornthanakasem W, Shotelersuk K, Termrungruanglert W, Voravud N, Niruthisard S, Mutirangura A. Human papillomavirus dna in plasma of patients with cervical cancer. *BMC Cancer* 2001;1(1):2.

[18] Shimada T, Yamaguchi N, Nishida N, Yamasaki K, Miura K, Katamine S, Masuzaki H. Human papillomavirus dna in plasma of patients with hpv16 dna-positive uterine cervical cancer. *Japanese Journal of Clinical Oncology* 2010;40(5):420 -424.

[19] Widschwendter A, Blassnig A, Wiedemair A, MÃ¼ller-Holzner E, MÃ¼ller HM, Marth C. Human papillomavirus dna in sera of cervical cancer patients as tumor marker. *Cancer Letters* 2003;202(2):231-239.

[20] Rombaldi RL, Serafini EP, Mandelli J, Zimmermann E, Losquiavo KP. Perinatal transmission of human papilomavirus dna. *Virol. J.* 2009;6:83.

[21] Weyn C, Thomas D, Jani J, Guizani M, Donner C, Van Rysselberge M, Hans C, Bossens M, Englert Y, Fontaine V. Evidence of human papillomavirus in the placenta. *J. Infect. Dis.* 2011;203(3):341-343.

[22] Foresta C, Garolla A, Zuccarello D, Pizzol D, Moretti A, Barzon L, Palù G. Human papillomavirus found in sperm head of young adult males affects the progressive motility. *Fertil. Steril.* 2010;93(3):802-806.

[23] Pittayakhajonwut D, Angeletti PC. Viral trans-factor independent replication of human papillomavirus genomes. *Virol. J.* 2010;7:123.

[24] Ahola H, Bergman P, Ström AC, Moreno-Lopéz J, Pettersson U. Organization and expression of the transforming region from the european elk papillomavirus (eepv). *Gene* 1986;50(1-3):195-205.

[25] DiMaio D, Mattoon D. Mechanisms of cell transformation by papillomavirus e5 proteins. *Oncogene* 2001;20(54):7866-7873.

[26] Schulz E, Gottschling M, Bravo IG, Wittstatt U, Stockfleth E, Nindl I. Genomic characterization of the first insectivoran papillomavirus reveals an unusually long, second non-coding region and indicates a close relationship to betapapillomavirus. *Journal of General Virology* 2009;90(3):626 -633.

[27] Nonnenmacher M, Salmon J, Jacob Y, Orth G, Breitburd F. Cottontail rabbit papillomavirus e8 protein is essential for wart formation and provides new insights into viral pathogenesis. *J. Virol.* 2006;80(10):4890-4900.

[28] Eriksson A, Stewart AC, Moreno-Lopéz J, Pettersson U. The genomes of the animal papillomaviruses european elk papillomavirus, deer papillomavirus, and reindeer papillomavirus contain a novel transforming gene (e9) near the early polyadenylation site. *J Virol.* 1994;68(12):8365-8373.

[29] Shah SD, Doorbar J, Goldstein RA. Analysis of host–parasite incongruence in papillomavirus evolution using importance sampling. *Molecular Biology and Evolution* 2010;27(6):1301 -1314.

[30] Gottschling M, Stamatakis A, Nindl I, Stockfleth E, Alonso Á, Bravo IG. Multiple evolutionary mechanisms drive papillomavirus diversification. *Molecular Biology and Evolution* 2007;24(5):1242 -1258.

[31] Shah SD, Doorbar J, Goldstein RA. Analysis of host–parasite incongruence in papillomavirus evolution using importance sampling. *Molecular Biology and Evolution* 2010;27(6):1301 -1314.

[32] Gottschling M, Göker M, Stamatakis A, Bininda-Emonds ORP, Nindl I, Bravo IG. Quantifying the phylodynamic forces driving papillomavirus evolution. *Molecular Biology and Evolution* 2011.

[33] Freitas AC. Recent insights into bovine papillomavirus. African Journal of Microbiology Research 2011;5(33).

[34] Batista MVA, Ferreira TAE, Freitas AC, Balbino VQ. An entropy-based approach for the identification of phylogenetically informative genomic regions of papillomavirus. *Infect. Genet. Evol.* 2011;11(8):2026-2033.

[35] Favre M, Ramoz N, Orth G. Human papillomaviruses: general features. *Clin. Dermatol.* 1997;15(2):181-198.

[36] Fehrmann F, Laimins LA. Human papillomaviruses: targeting differentiating epithelial cells for malignant transformation. *Oncogene* 2003;22(33):5201-5207.

[37] McCance DJ. Transcriptional regulation by human papillomaviruses. *Curr. Opin. Genet. Dev.* 2005;15(5):515-519.

[38] Stubenrauch F, Laimins LA. Human papillomavirus life cycle: active and latent phases. Semin. Cancer Biol. 1999;9(6):379-386.

[39] Stanley M. Immune responses to human papillomavirus. *Vaccine* 2006;24 Suppl 1:S16-22.

[40] Zhang P, Nouri M, Brandsma JL, Iftner T, Steinberg BM. Induction of e6/e7 expression in cottontail rabbit papillomavirus latency following uv activation. *Virology* 1999;263(2):388-394.

[41] zur Hausen H. Papillomavirus infections--a major cause of human cancers. *Biochim. Biophys. Acta* 1996;1288(2):F55-78.

[42] Campo MS. Animal models of papillomavirus pathogenesis. *Virus Res.* 2002;89(2):249-261.

[43] Schiffman M, Rodriguez AC, Chen Z, Wacholder S, Herrero R, Hildesheim A, Desalle R, Befano B, Yu K, Safaeian M, Sherman ME, Morales J, Guillen D, Alfaro M, Hutchinson M, Solomon D, Castle PE, Burk RD. A population-based prospective study of carcinogenic human papillomavirus variant lineages, viral persistence, and cervical neoplasia. *Cancer Research* 2010;70(8):3159 -3169.

[44] Doorbar J. Papillomavirus life cycle organization and biomarker selection. *Dis. Markers* 2007;23(4):297-313.

[45] Bruni L, Diaz M, Castellsagué M, Ferrer E, Bosch FX, de Sanjosé S. Cervical human papillomavirus prevalence in 5 continents: meta-analysis of 1 million women with normal cytological findings. Journal of Infectious Diseases 2010;202(12):1789 -1799.

[46] Woodman CBJ, Collins SI, Young LS. The natural history of cervical hpv infection: unresolved issues. *Nat Rev Cancer* 2007;7(1):11-22.

[47] Li N, Franceschi S, Howell-Jones R, Snijders PJF, Clifford GM. Human papillomavirus type distribution in 30,848 invasive cervical cancers worldwide: variation by geographical region, histological type and year of publication. International *Journal of Cancer* 2011;128(4):927-935.

[48] Broso PR, Buffetti G. [the papanicolaou classification in the bethesda system (national cancer institute, bethesda, maryland)]. *Minerva Ginecol* 1993;45(11):557-563.

[49] Trimble CL, Piantadosi S, Gravitt P, Ronnett B, Pizer E, Elko A, Wilgus B, Yutzy W, Daniel R, Shah K, Peng S, Hung C, Roden R, Wu TC, Pardoll D, The Johns Hopkins Medical Institutions B. Spontaneous regression of high-grade cervical dysplasia: effects of human papillomavirus type and hla phenotype. *Clinical Cancer Research* 2005;11(13):4717 -4723.

[50] Pett MR, Alazawi WOF, Roberts I, Dowen S, Smith DI, Stanley MA, Coleman N. Acquisition of high-level chromosomal instability is associated with integration of human papillomavirus type 16 in cervical keratinocytes. *Cancer Res.* 2004;64(4):1359-1368.

[51] Steben M, Duarte-Franco E. Human papillomavirus infection: epidemiology and pathophysiology. *Gynecol. Oncol.* 2007;107(2 Suppl 1):S2-5.

[52] Bloch N, Sutton RH, Breen M, Spradbrow PB. Bovine papillomavirus type 4 in australia. *Aust. Vet. J.* 1995;72(7):273-274.

[53] Ziegert C, Wentzensen N, Vinokurova S, Kisseljov F, Einenkel J, Hoeckel M, von Knebel Doeberitz M. A comprehensive analysis of hpv integration loci in anogenital lesions combining transcript and genome-based amplification techniques. *Oncogene* 2003;22(25):3977-3984.

[54] WHO/ICO Information Centre on HPV and Cervical Cancer (HPV Information Centre). Human papillomavirus and related cancers in Brazil. 2010.

[55] Baldez da Silva MFPT, Chagas BS, Guimarães V, Katz LMC, Felix PM, Miranda PM, Lima AA, Arraes LC, Martins DBG, Lima Filho JL, Stocco RC, Crovella S, Freitas AC, Beçak W. Hpv31 and hpv33 incidence in cervical samples from women in recife, Brazil. *Genet. Mol. Res.* 2009;8(4):1437-1443.

[56] Cerqueira DM, Raiol T, Véras NMC, Gal Milanezi N, Amaral FA, Macedo Brígido M, Martins CRF. New variants of human papillomavirus type 18 identified in central Brazil. *Virus Genes* 2008;37:282-287.

[57] Stewart AC, Eriksson AM, Manos MM, Muñoz N, Bosch FX, Peto J, Wheeler CM. Intratype variation in 12 human papillomavirus types: a worldwide perspective. *J. Virol.* 1996;70(5):3127-3136.

[58] Calleja-Macias IE, Villa LL, Prado JC, Kalantari M, Allan B, Williamson A-L, Chung L-P, Collins RJ, Zuna RE, Dunn ST, Chu T-Y, Cubie HA, Cuschieri K, von Knebel-Doeberitz M, Martins CR, Sanchez GI, Bosch FX, Munoz N, Bernard H-U. Worldwide

genomic diversity of the high-risk human papillomavirus types 31, 35, 52, and 58, four close relatives of human papillomavirus type 16. *J. Virol.* 2005;79(21):13630-13640.

[59] Prado JC, Calleja-Macias IE, Bernard H-U, Kalantari M, Macay SA, Allan B, Williamson A-L, Chung L-P, Collins RJ, Zuna RE, Dunn ST, Ortiz-Lopez R, Barrera-Saldaña HA, Cubie HA, Cuschieri K, von Knebel-Doeberitz M, Sanchez GI, Bosch FX, Villa LL. Worldwide genomic diversity of the human papillomaviruses-53, 56, and 66, a group of high-risk hpvs unrelated to hpv-16 and hpv-18. *Virolog*y 2005;340(1):95-104.

[60] Xi LF, Critchlow CW, Wheeler CM, Koutsky LA, Galloway DA, Kuypers J, Hughes JP, Hawes SE, Surawicz C, Goldbaum G, Holmes KK, Kiviat NB. Risk of anal carcinoma in situ in relation to human papillomavirus type 16 variants. *Nat Cancer Res.* 1998;58(17):3839-3844.

[61] Villa LL, Sichero L, Rahal P, Caballero O, Ferenczy A, Rohan T, Franco EL. Molecular variants of human papillomavirus types 16 and 18 preferentially associated with cervical neoplasia. *Nat J. Gen. Virol.* 2000;81(Pt 12):2959-2968.

[62] Lizano M, Guido MC, García-Carrancá A, Berumen J, Casas L. Association between human papillomavirus type 18 variants and histopathology of cervical cancer. *Journal of the National Cancer Institute* 1997;89(16):1227 -1231.

[63] Calleja-Macias IE, Kalantari M, Huh J, Ortiz-Lopez R, Rojas-Martinez A, Gonzalez-Guerrero JF, Williamson A-L, Hagmar B, Wiley DJ, Villarreal L, Bernard H-U, Barrera-Saldaña HA. Genomic diversity of human papillomavirus-16, 18, 31, and 35 isolates in a mexican population and relationship to european, african, and native american variants. *Virolog*y 2004;319(2):315-323.

[64] Ho L, Chan SY, Burk RD, Das BC, Fujinaga K, Icenogle JP, Kahn T, Kiviat N, Lancaster W, Mavromara-Nazos P. The genetic drift of human papillomavirus type 16 is a means of reconstructing prehistoric viral spread and the movement of ancient human populations. *Nat J Virol* 1993;67(11):6413-6423.

[65] Yamada T, Wheeler CM, Halpern AL, Stewart AC, Hildesheim A, Jenison SA. Human papillomavirus type 16 variant lineages in united states populations characterized by nucleotide sequence analysis of the e6, l2, and l1 coding segments. *J. Virol.* 1995;69(12):7743-7753.

[66] Yamada T, Manos MM, Peto J, Greer CE, Munoz N, Bosch FX, Wheeler CM. Human papillomavirus type 16 sequence variation in cervical cancers: a worldwide perspective. *Nat J Virol* 1997;71(3):2463-2472.

[67] Ong CK, Chan SY, Campo MS, Fujinaga K, Mavromara-Nazos P, Labropoulou V, Pfister H, Tay SK, ter Meulen J, Villa LL. Evolution of human papillomavirus type 18: an ancient phylogenetic root in africa and intratype diversity reflect coevolution with human ethnic groups. *Nat J Virol* 1993;67(11):6424-6431.

[68] Arias-Pulido H, Peyton CL, Torrez-Martínez N, Anderson DN, Wheeler CM. Human papillomavirus type 18 variant lineages in united states populations characterized by sequence analysis of lcr-e6, e2, and l1 regions. *Virolog*y 2005;338(1):22-34.

[69] Bernard H-U. The clinical importance of the nomenclature, evolution and taxonomy of human papillomaviruses. *J. Clin. Virol.* 2005;32 Suppl 1:S1-6.

[70] Chagas BS, Batista MVA, Guimarães V, Balbino VQ, Crovella S, Freitas AC. New variants of e6 and e7 oncogenes of human papillomavirus type 31 identified in northeastern brazil. *Nat Gynecol. Oncol.* 2011;123(2):284-288.

[71] Franco EL, Rohan TE, Villa LL. Epidemiologic evidence and human papillomavirus infection as a necessary cause of cervical cancer. *Journal of the National Cancer Institute* 1999;91(6):506 -511.

[72] Bernard H-U. Gene expression of genital human papillomaviruses and considerations on potential antiviral approaches. *Antivir. Ther.* (Lond.) 2002;7(4):219-237.

[73] Sichero L, Villa LL. Epidemiological and functional implications of molecular variants of human papillomavirus. *Braz. J. Med. Biol. Res.* 2006;39(6):707-717.

[74] Kast W, Brandt R, Sidney J, Drijfhout J, Kubo R, Grey H, Melief C, Sette A. Role of hla-a motifs in identification of potential ctl epitopes in human papillomavirus type 16 e6 and e7 proteins. *The Journal of Immunology* 1994;152(8):3904 -3912.

[75] Ho L, Chan SY, Burk RD, Das BC, Fujinaga K, Icenogle JP, Kahn T, Kiviat N, Lancaster W, Mavromara-Nazos P. The genetic drift of human papillomavirus type 16 is a means of reconstructing prehistoric viral spread and the movement of ancient human populations. *J. Virol.* 1993;67(11):6413-6423.

[76] Ong CK, Chan SY, Campo MS, Fujinaga K, Mavromara-Nazos P, Labropoulou V, Pfister H, Tay SK, ter Meulen J, Villa LL. Evolution of human papillomavirus type 18: an ancient phylogenetic root in africa and intratype diversity reflect coevolution with human ethnic groups. *J. Virol.* 1993;67(11):6424-6431.

[77] Xi LF, Demers GW, Koutsky LA, Kiviat NB, Kuypers J, Watts DH, Holmes KK, Galloway DA. Analysis of human papillomavirus type 16 variants indicates establishment of persistent infection. *The Journal of Infectious Diseases* 1995;172(3):747-755.

[78] Londesborough P, Ho L, Terry G, Cuzick J, Wheeler C, Singer A. Human papillomavirus genotype as a predictor of persistence and development of high-grade lesions in women with minor cervical abnormalities. *Int. J. Cancer* 1996;69(5):364-368.

[79] Xi LF, Koutsky LA, Galloway DA, Kiviat NB, Kuypers J, Hughes JP, Wheeler CM, Holmes KK. Genomic variation of human papillomavirus type 16 and risk for high grade cervical intraepithelial neoplasia. *Nat Journal of the National Cancer Institute* 1997;89(11):796 -802.

[80] Zehbe I, Wilander E, Delius H, Tommasino M. Human papillomavirus 16 e6 variants are more prevalent in invasive cervical carcinoma than the prototype. *Nat Cancer Res.* 1998;58(4):829-833.

[81] Nindl I, Rindfleisch K, Lotz B, Schneider A, Dürst M. Uniform distribution of hpv 16 e6 and e7 variants in patients with normal histology, cervical intra-epithelial neoplasia and cervical cancer. *Nat International Journal of Cancer* 1999;82(2):203-207.

[82] Hildesheim A, Schiffman M, Bromley C, Wacholder S, Herrero R, Rodriguez AC, Bratti MC, Sherman ME, Scarpidis U, Lin Q-Q, Terai M, Bromley RL, Buetow K, Apple RJ, Burk RD. Human papillomavirus type 16 variants and risk of cervical cancer. *Nat Journal of the National Cancer Institute* 2001;93(4):315 -318.

[83] Zehbe I, Voglino G, Wilander E, Delius H, Marongiu A, Edler L, Klimek F, Andersson S, Tommasino M. P53 codon 72 polymorphism and various human papillomavirus 16 e6 genotypes are risk factors for cervical cancer development. *Nat Cancer Res.* 2001;61(2):608-611.

[84] Villa LL, Sichero L, Rahal P, Caballero O, Ferenczy A, Rohan T, Franco EL. Molecular variants of human papillomavirus types 16 and 18 preferentially associated with cervical neoplasia. *Nat J. Gen. Virol.* 2000;81(Pt 12):2959-2968.

[85] Xi LF, Kiviat NB, Hildesheim A, Galloway DA, Wheeler CM, Ho J, Koutsky LA. Human papillomavirus type 16 and 18 variants: race-related distribution and persistence. *J. Natl. Cancer Inst.* 2006;98(15):1045-1052.

[86] Sichero L, Ferreira S, Trottier H, Duarte-Franco E, Ferenczy A, Franco EL, Villa LL. High grade cervical lesions are caused preferentially by non-european variants of hpvs 16 and 18. *Int. J. Cancer* 2007;120(8):1763-1768.

[87] Kämmer C, Tommasino M, Syrjänen S, Delius H, Hebling U, Warthorst U, Pfister H, Zehbe I. Variants of the long control region and the e6 oncogene in european human papillomavirus type 16 isolates: implications for cervical disease. *Br. J. Cancer* 2002;86(2):269-273.

[88] de Boer MA, Peters LAW, Aziz MF, Siregar B, Cornain S, Vrede MA, Jordanova ES, Kolkman-Uljee S, Fleuren GJ. Human papillomavirus type 16 e6, e7, and l1 variants in cervical cancer in indonesia, suriname, and the netherlands. *Nat Gynecol. Oncol.* 2004;94(2):488-494.

[89] Matsumoto K, Yoshikawa H, Nakagawa S, Tang X, Yasugi T, Kawana K, Sekiya S, Hirai Y, Kukimoto I, Kanda T, Taketani Y. Enhanced oncogenicity of human papillomavirus type 16 (hpv16) variants in japanese population. *Cancer Lett.* 2000;156(2):159-165.

[90] Hu X, Pang T, Guo Z, Mazurenko N, Kisseljov F, Pontén J, Nistér M. Hpv16 e6 gene variations in invasive cervical squamous cell carcinoma and cancer in situ from russian patients. *Br J Cancer* 2001;84(6):791-795.

[91] Song YS, Kee SH, Kim JW, Park NH, Kang SB, Chang WH, Lee HP. Major sequence variants in e7 gene of human papillomavirus type 16 from cervical cancerous and noncancerous lesions of korean women. *Gynecologic Oncology* 1997;66(2):275-281.

[92] De Boer MA, Peters LAW, Aziz MF, Siregar B, Cornain S, Vrede MA, Jordanova ES, Fleuren GJ. Human papillomavirus type 18 variants: histopathology and e6/e7 polymorphisms in three countries. *Int. J. Cancer* 2005;114(3):422-425.

[93] De la Cruz-Hernández E, García-Carrancá A, Mohar-Betancourt A, Dueñas-González A, Contreras-Paredes A, Pérez-Cardenas E, Herrera-Goepfert R, Lizano-Soberón M. Differential splicing of e6 within human papillomavirus type 18 variants and functional consequences. *Journal of General Virology* 2005;86(9):2459 -2468.

[94] Rose BR, Thompson CH, Zhang J, Stoeter M, Stephen A, Pfister H, Tattersall MH, Cossart YE. Sequence variation in the upstream regulatory region of hpv 18 isolates from cervical cancers. *Nat Gynecol. Oncol.* 1997;66(2):282-289.

[95] Combita A-L, Touze A, Bousarghin L, Christensen ND, Coursaget P. Identification of two cross-neutralizing linear epitopes within the l1 major capsid protein of human papillomaviruses. *J. Virol.* 2002;76(13):6480-6486.

[96] Kirnbauer R, Taub J, Greenstone H, Roden R, Dürst M, Gissmann L, Lowy DR, Schiller JT. Efficient self-assembly of human papillomavirus type 16 l1 and l1-l2 into virus-like particles. *J. Virol.* 1993;67(12):6929-6936.

[97] Bishop B, Dasgupta J, Chen XS. Structure-based engineering of papillomavirus major capsid l1: controlling particle assembly. *Virology Journal* 2007;4(1):3.

[98] Xi LF, Koutsky LA, Hildesheim A, Galloway DA, Wheeler CM, Winer RL, Ho J, Kiviat NB. Risk for high-grade cervical intraepithelial neoplasia associated with variants of human papillomavirus types 16 and 18. Cancer Epidemiol. *Biomarkers Prev*. 2007;16(1):4-10.

[99] De Boer MA, Peters LAW, Aziz MF, Siregar B, Cornain S, Vrede MA, Jordanova ES, Fleuren GJ. Human papillomavirus type 18 variants: histopathology and e6/e7 polymorphisms in three countries. *Int. J. Cancer* 2005;114(3):422-425.

[100] Raiol T, Wyant PS, de Amorim RMS, Cerqueira DM, Milanezi N von G, Brígido M de M, Sichero L, Martins CRF. Genetic variability and phylogeny of the high-risk hpv-31, -33, -35, -52, and -58 in central brazil. *Journal of Medical Virology* 2009;81:685-692.

[101] Comerford SA, McCance DJ, Dougan G, Tite JP. Identification of t- and b-cell epitopes of the e7 protein of human papillomavirus type 16. *J. Virol.* 1991;65(9):4681-4690.

[102] Cento V, Rahmatalla N, Ciccozzi M, Perno CF, Ciotti M. Intratype variations of hpv 31 and 58 in italian women with abnormal cervical cytology. *J. Med. Virol.* 2011;83(10):1752-1761.

[103] Muñoz N. Human papillomavirus and cancer: the epidemiological evidence. J. Clin. Virol. 2000;19(1-2):1-5.

[104] Forslund O. Genetic diversity of cutaneous human papillomaviruses. *Nat J. Gen. Virol.* 2007;88(Pt 10):2662-2669.

[105] Feltkamp MCW, de Koning MNC, Bavinck JNB, Ter Schegget J. Betapapillomaviruses: innocent bystanders or causes of skin cancer. *J. Clin. Virol.* 2008;43(4):353-360.

[106] United States Department of Agriculture (USDA). Livestock and poultry: world market and trades. 2011.

[107] Food and Agricultural Organization (FAO). World cow's milk production. 2010.

[108] Borzacchiello G, Roperto F. Bovine papillomaviruses, papillomas and cancer in cattle. Veterinary Research 2008;39(5):19.

[109] Leal AM, Ferraz OP, Carvalho C, Freitas AC, Beniston RG, Beçak W, Campo MS, Stocco dos Santos RC. Quercetin induces structural chromosomal aberrations and uncommon rearrangements in bovine cells transformed by the e7 protein of bovine papillomavirus type 4. *Vet Comp Oncol* 2003;1(1):15-21.

[110] Hatama S, Ishihara R, Ueda Y, Kanno T, Uchida I. Detection of a novel bovine papillomavirus type 11 (bpv-11) using xipapillomavirus consensus polymerase chain reaction primers. *Arch. Virol.* 2011;156(7):1281-1285.

[111] Hatama S, Nobumoto K, Kanno T. Genomic and phylogenetic analysis of two novel bovine papillomaviruses, bpv-9 and bpv-10. *Journal of General Virology* 2008;89(1):158 -163.

[112] Zhu W, Dong J, Shimizu E, Hatama S, Kadota K, Goto Y, Haga T. Characterization of novel bovine papillomavirus type 12 (bpv-12) causing epithelial papilloma. *Nat Arch. Virol.* 2012;157(1):85-91.

[113] Löhr CV, Juan-Sallés C, Rosas-Rosas A, García AP, Garner MM, Teifke JP. Sarcoids in captive zebras (equus burchellii): association with bovine papillomavirus type 1 infection. *Journal of Zoo and Wildlife Medicine* 2005;36(1):74-81.

[114] Kidney BA, Berrocal A. Sarcoids in two captive tapirs (tapirus bairdii): clinical, pathological and molecular study. *Vet. Dermatol.* 2008;19(6):380-384.

[115] Silvestre O, Borzacchiello G, Nava D, Iovane G, Russo V, Vecchio D, D'Ausilio F, Gault EA, Campo MS, Paciello O. Bovine papillomavirus type 1 dna and e5 oncoprotein expression in water buffalo fibropapillomas. *Veterinary Pathology Online* 2009;46(4):636 -641.
[116] van Dyk E, Oosthuizen MC, Bosman A-M, Nel PJ, Zimmerman D, Venter EH. Detection of bovine papillomavirus dna in sarcoid-affected and healthy free-roaming zebra (equus zebra) populations in south africa. *Journal of Virological Methods* 2009;158(1–2):141-151.
[117] Pangty K, Singh S, Goswami R, Saikumar G, Somvanshi R. Detection of bpv-1 and -2 and quantification of bpv-1 by real-time pcr in cutaneous warts in cattle and buffaloes. *Transboundary and Emerging Diseases* 2010;57(3):185-196.
[118] van Dyk E, Bosman AM, van Wilpe E, Williams JH, Bengis RG, van Heerden J, Venter EH. Detection and characterisation of papillomavirus in skin lesions of giraffe and sable antelope in south africa. *J S Afr Vet Assoc* 2011;82(2):80-85.
[119] Campo MS. Bovine papillomavirus: old system, new lessons? 2006.
[120] MacLaughlan SD, Lachance JA, Gjelsvik A. Correlation between smoking status and cervical cancer screening: a cross-sectional study. *J Low Genit Tract Dis* 2011;15(2):114-119.
[121] Olson C, Olson RO, Hubbard-Van Stelle S. Variations of response of cattle to experimentally induced viral papillomatosis. *J. Am. Vet. Med. Assoc.* 1992;201(1):56-62.
[122] Hama, C, Matsumoto, T, Franceschini, P. H. Papilomatose bovina: avaliação clínica de diferentes produtos utilizados no controle e tratamento., vol. 2. *Ciência Veterinária,* [s.d.].
[123] Melo. C. B, Leite, R. C. Papilomatose bovina, vol. 6. Ciência Veterinária nos Trópicos, [s.d.].
[124] Muro, L. F. F., Bottura, C. R. P., Piccinin, A. *Papilomatose bovina.*, vol. 1. [s.d.].
[125] Mbuthia, G. P., Ngatia, T. A., & Wamokoya, J. P. O. Occurrence of bovine skin diseases in kenya., vol. 41. *Bulletin of Animal Production in Africa*, [s.d.].
[126] Campo MS, Jarrett WF, O'Neil W, Barron RJ. Latent papillomavirus infection in cattle. *Res. Vet. Sci.* 1994;56(2):151-157.
[127] Pfister H, Linz U, Gissmann L, Huchthausen B, Hoffmann D, Zur Hausen H. Partial characterization of a new type of bovine papilloma viruses. *Virology* 1979;96(1):1-8.
[128] Carvalho CCR, Batista MVA, Silva MAR, Balbino VQ, Freitas AC. Detection of bovine papillomavirus types, co-infection and a putative new bpv11 subtype in cattle. *Transboundary and Emerging Diseases* 2012.
[129] Freitas AC de, Carvalho C de, Brunner O, Birgel-Junior EH, Dellalibera AMMP, Benesi FJ, Gregory L, Beçak W, Santos R de CS dos. Viral dna sequences in peripheral blood and vertical transmission of the virus: a discussion about bpv-1. *Brazilian Journal of Microbiology* 2003;34:76-78.
[130] Freitas, A. C., Silva, M. A. R., Carvalho, C. C. R., Birgel Jr, E. H., Santos, J. F., Beçak, W., & Stocco dos Santos, R. C. Papillomavirus dna detection in non epithelial tissues: a discussion about bovine papillomavirus. in a. méndez-vilas (ed.), formatex microbiology book series. *Mérida, EX: Formatex*, [s.d.].
[131] Stocco R de C. Papillomatosis in cattle: in situ detection of bovine papillomavirus dna sequences in reproductive tissues [s.d.].

[132] Brandt S, Haralambus R, Schoster A, Kirnbauer R, Stanek C. Peripheral blood mononuclear cells represent a reservoir of bovine papillomavirus dna in sarcoid-affected equines. *Journal of General Virology* 2008;89(6):1390 -1395.

[133] Roperto S, Brun R, Paolini F, Urraro C, Russo V, Borzacchiello G, Pagnini U, Raso C, Rizzo C, Roperto F, Venuti A. Detection of bovine papillomavirus type 2 in the peripheral blood of cattle with urinary bladder tumours: possible biological role. *Journal of General Virology* 2008;89(12):3027 -3033.

[134] Yaguiu A, Dagli MLZ, Birgel EH Jr, Alves Reis BCA, Ferraz OP, Pituco EM, Freitas AC, Beçak W, Stocco RC. Simultaneous presence of bovine papillomavirus and bovine leukemia virus in different bovine tissues: in situ hybridization and cytogenetic analysis. *Genet. Mol. Res.* 2008;7(2):487-497.

[135] Roperto S, Comazzi S, Ciusani E, Paolini F, Borzacchiello G, Esposito I, Lucà R, Russo V, Urraro C, Venuti A, Roperto F. Pbmcs are additional sites of productive infection of bovine papillomavirus type 2. *Journal of General Virology* 2011;92(8):1787 -1794.

[136] Carvalho C de, Freitas AC de, Brunner O, Góes LGB, Cavalcante AY, Beçak W, Santos R de CS dos. Bovine papillomavirus type 2 in reproductive tract and gametes of slaughtered bovine females. *Braz. J. Microbiol* 2003;34(supl.1):82-84.

[137] Lindsey CL, Almeida ME, Vicari CF, Carvalho C, Yaguiu A, Freitas AC, Beçak W, Stocco RC. Bovine papillomavirus dna in milk, blood, urine, semen, and spermatozoa of bovine papillomavirus-infected animals. *Nat Genet. Mol. Res.* 2009;8(1):310-318.

[138] Silva MAR, Pontes NE, Da Silva KMG, Guerra MMP, Freitas AC. Detection of bovine papillomavirus type 2 dna in commercial frozen semen of bulls (bos taurus). *Anim. Reprod. Sci.* 2011;129(3-4):146-151.

[139] Stocco dos Santos RC, Lindsey CJ, Ferraz OP, Pinto JR, Mirandola RS, Benesi FJ, Birgel EH, Pereira CA, Be√ßak W. Bovine papillomavirus transmission and chromosomal aberrations: an experimental model. *Journal of General Virology* 1998;79(9):2127 -2135.

[140] Schmitt M, Fiedler V, Müller M. Prevalence of bpv genotypes in a german cowshed determined by a novel multiplex bpv genotyping assay. *Journal of Virological Methods* 2010;170(1–2):67-72.

[141] Maeda Y, Shibahara T, Wada Y, Kadota K, Kanno T, Uchida I, Hatama S. An outbreak of teat papillomatosis in cattle caused by bovine papilloma virus (bpv) type 6 and unclassified bpvs. *Veterinary Microbiology* 2007;121(3–4):242-248.

[142] Ogawa T, Tomita Y, Okada M, Shinozaki K, Kubonoya H, Kaiho I, Shirasawa H. Broad-spectrum detection of papillomaviruses in bovine teat papillomas and healthy teat skin. *Journal of General Virology* 2004;85(8):2191 -2197.

[143] Tomita Y, Literak I, Ogawa T, Jin Z, Shirasawa H. Complete genomes and phylogenetic positions of bovine papillomavirus type 8 and a variant type from a european bison. *Virus Genes* 2007;35(2):243-249.

[144] Singh V. Papillomatosis in indian cattle: occurrence and etiopathology. *Indian Journal of Veterinary Pathology* [s.d.];33(1):52-57.

[145] Rai GK, Saxena M, Singh V, Somvanshi R, Sharma B. Identification of bovine papilloma virus 10 in teat warts of cattle by dnase-sispa. *Veterinary Microbiology* 2011;147(3–4):416-419.

[146] Carr EA, Théon AP, Madewell BR, Griffey SM, Hitchcock ME. Bovine papillomavirus dna in neoplastic and nonneoplastic tissues obtained from horses with and without sarcoids in the western united states. *Nat Am. J. Vet. Res.* 2001;62(5):741-744.

[147] Wobeser BK, Davies JL, Hill JE, Jackson ML, Kidney BA, Mayer MN, Townsend HGG, Allen AL. Epidemiology of equine sarcoids in horses in western canada. *Can Vet J* 2010;51(10):1103-1108.

[148] Spradbrow PB, Ford J. Bovine papillomavirus type 2 from bovine cutaneous papillomas in australia. *Australian Veterinary Journal* 1983;60(3):78-79.

[149] Silva MS e, Weiss M, Brum MCS, dos Anjos BL, Torres FD, Weiblen R, Flores EF. Molecular identification of bovine papillomaviruses associated with cutaneous warts in southern Brazil. *Journal of Veterinary Diagnostic Investigation* 2010;22(4):603 -606.

[150] Claus MP, Vivian D, Lunardi M, Alfieri AF, Alfieri AA. Phylogenetic analysis of bovine papillomavirus associated with skin warts in cattle herds from the state of paraná. *Pesquisa Veterinária Brasileira* 2007;27(7):314-318.

[151] Claus MP, Lunardi M, Alfieri AF, Ferracin LM, Fungaro MHP, Alfieri AA. Identification of unreported putative new bovine papillomavirus types in brazilian cattle herds. *Vet. Microbiol.* 2008;132(3-4):396-401.

[152] Wosiacki SR, Claus MP, Alfieri AF, Alfieri AA. Bovine papillomavirus type 2 detection in the urinary bladder of cattle with chronic enzootic haematuria. Mem. Inst. Oswaldo Cruz 2006;101(6):635-638.

[153] Diniz N, Melo TC, Santos JF, Mori E, Brandão PE, Richtzenhain LJ, Freitas AC, Beçak W, Carvalho RF, Stocco RC. Simultaneous presence of bovine papillomavirus in blood and in short-term lymphocyte cultures from dairy cattle in pernambuco, brazil. *Nat Genet. Mol. Res.* 2009;8(4):1474-1480.

[154] Jarrett WFH, McNeil PE, Grimshaw WTR, Selman IE, McIntyre WIM. High incidence area of cattle cancer with a possible interaction between an environmental carcinogen and a papilloma virus. *Nature* 1978;274(5668):215-217.

[155] Borzacchiello G, Ambrosio V, Roperto S, Poggiali F, Tsirimonakis E, Venuti A, Campo MS, Roperto F. Bovine papillomavirus type 4 in oesophageal papillomas of cattle from the south of Italy. *J. Comp. Pathol.* 2003;128(2-3):203-206.

[156] Leishangthem GD. Detection of bovine papillomaviruses in cutaneous warts/papillomas in cattle. *Indian Journal of Veterinary Pathology* [s.d.];32(1):15-20.

[157] Claus MP, Lunardi M, Alfieri AA, Otonel RAA, Sartori D, Fungaro MHP, Alfieri AF. Multiple bovine papillomavirus infections associated with cutaneous papillomatosis in brazilian cattle herds. *Braz. arch. biol. technol* 2009;52(spe):93-98.

[158] Antonsson A, Hansson BG. Healthy skin of many animal species harbors papillomaviruses which are closely related to their human counterparts. *Journal of Virology* 2002;76(24):12537 -12542.

[159] Giannoudis A, Herrington CS. Human papillomavirus variants and squamous neoplasia of the cervix. *J. Pathol.* 2001;193(3):295-302.

[160] Reid S, Smith K, Jarrett W. Detection, cloning and characterisation of papillomaviral dna present in sarcoid tumours of equus asinus. *Veterinary Record* 1994;135(18):430 - 432.

[161] Chambers G, Ellsmore VA, O'Brien PM, Reid SWJ, Love S, Campo MS, Nasir L. Sequence variants of bovine papillomavirus e5 detected in equine sarcoids. *Virus Research* 2003;96(1–2):141-145.

[162] Nasir L, Gault E, Morgan IM, Chambers G, Ellsmore V, Campo MS. Identification and functional analysis of sequence variants in the long control region and the e2 open reading frame of bovine papillomavirus type 1 isolated from equine sarcoids. *Virology* 2007;364(2):355-361.

[163] Szczerba-Turek A, Siemionek J, Wasowicz K, Szweda W, Raś A, Platt-Samoraj A. Partial sequence analysis of the l1 gene of bovine papillomavirus type 1 detected by pcr with my09/my11 primers in equine sarcoids in poland. *Pol J Vet Sci* 2010;13(2):241-246.

[164] Freitas AC. Recent insights into bovine papillomavirus. *African Journal of Microbiology Research* 2011;5(33).

[165] García-Vallvé S, Alonso A, Bravo IG. Papillomaviruses: different genes have different histories. *Trends Microbiol.* 2005;13(11):514-521.

[166] Bogaert L, Martens A, Kast WM, Van Marck E, De Cock H. Bovine papillomavirus dna can be detected in keratinocytes of equine sarcoid tumors. *Vet. Microbiol.* 2010;146(3-4):269-275.

[167]Turner PE, Elena SF. Cost of host radiation in an rna virus. *Genetics* 2000;156(4):1465-1470.

In: Advances in Genetics Research. Volume 12
Editor: Kevin V. Urbano

ISBN: 978-1-63321-351-7

Chapter 5

TELOMERE LENGTH IN LYMPHOID MALIGNANCIES: A COMPARATIVE ANALYSIS

Julieta Panero, Patricia Dos Santos and Irma Slavutsky*
Departamento de Genética, Instituto de Investigaciones Hematológicas "Mariano R. Castex", Academia Nacional de Medicina, Buenos Aires, Argentina

ABSTRACT

Human telomeres are nucleoprotein complexes located at the end of chromosomes, composed of tandem repeats of the non-codificant DNA sequence TTAGGG. A core of six telomere binding proteins, the shelterin complex, serves to protect telomeric ends and to prevent recognition of chromosome ends as damaged DNA. Both critically short telomeres and direct disruption of the shelterin structure can initiate telomere dysfunction. The enzyme telomerase is a ribonucleoprotein that compensates the telomere reduction by adding new repeats to chromosome ends. Different studies have demonstrated that reduced telomere length (TL) contribute directly to genomic instability, usually found in several types of cancer. Short telomeres and high telomerase activity have been associated with tumorigenesis in patients with hematological neoplasm. Lymphoid malignancies constitute a heterogeneous group of lymphoproliferative disorders that represent distinct disease entities, according to their clinical, morphological, immunophenotypic and genetic features. In this study, we analyzed the TL in 262 patients with different B-cell malignancies and its association with genetic characteristics and telomere-associated proteins expression. TL was highly heterogeneous and significantly reduced in B-cell disorders compared to normal controls. Patients with diffuse large B-cell lymphoma secondary to follicular lymphoma (DLBCL-S) had the shortest telomeres, followed by follicular lymphoma (FL). Mantle cell lymphoma (MCL) and *de novo* DLBCL showed similar mean TL, whereas monoclonal gammopathy of undetermined significance (MGUS) and chronic lymphocytic leukemia (CLL) cases displayed the longest TL. Mean TL for multiple myeloma (MM) were intermediate among those entities. Significant shorter TL in MM and CLL patients with chromosomal abnormalities with respect to cases with normal karyotypes were observed

* Correspondence to: Irma Slavutsky, MD, PhD. Departamento de Genética, Instituto de Investigaciones Hematológicas "Mariano R. Castex". Academia Nacional de Medicina Pacheco de Melo 3081. 1425 - Ciudad Autónoma de Buenos Aires, Argentina,. FAX: (5411) 4803-947. Email: islavutsky@hematologia.anm.edu.ar

($p<0.03$). In FL, a significant difference between *BCL2* positive and negative cases was found ($p<0.05$). The analysis of telomere-associated proteins expression was performed in MGUS and MM. A significant increase of *TRF1* levels and a lower expression of *TRF2* and *TANK1* in MGUS with respect to MM patients were found ($p=0.006$). The analysis of TL according to *hTERT* expression showed that patients with the highest mRNA values had the shortest TL. These findings are of interest considering the new therapeutic approaches with diverse telomere-targeting inhibitors and suggest that a number of patients with B-cell malignancies may be particularly sensitive to telomere-damage drugs.

INTRODUCTION

Telomeres

Human telomeres are nucleoprotein complexes located at the end of chromosomes, composed of tandem repeats of the non-codificant DNA sequence TTAGGG. They act as protective caps, preventing both degradation of chromosome ends and their recognition as double strand breaks, which might otherwise result in end-to-end fusions and aberrant recombination [1]. The maintenance of chromosomal stability and genome integrity is mainly due to DNA-binding proteins, which associate with telomeric DNA and other proteins/complexes to achieve telomere-end protection and length control. As chromosomes are replicated during cell division, there is an incomplete replication of the 3' end of the lagging strand. This is known as the "end replication problem" and results in the loss of approximately 50 bp of telomeric sequence with each cell division, leading to telomere dysfunction and, ultimately, contributing to tumorigenesis. Analysis of telomeres in different organisms revealed that 3' single-stranded overhang can invade the double-stranded region, displacing the homologous strand of the same telomere [2]. Consequently, the telomere forms a protective structure termed t-loop, which has a crucial role in sequestering the 3' end of telomeric DNA, preventing recognition by the DNA damage machinery.

The length of the telomeric DNA varies among species, being human telomeres around 10-15 kbp in length. It is known that cells can undergo only a limited number of divisions, known as the Hayflick limit, before entering senescence where they remain metabolically active but have lost their replicative capacity [2, 3]. Reduction in telomere length could provide the signal to cause growth arrest. *In vitro*, cells can be induced to continue dividing beyond the Hayflick limit by expression of some oncogenes [4]. During this process, their telomeres continue to shorten with each cell division until they enter to crisis where the majority will die. Some immortalized cells that emerge from crisis express the enzyme telomerase. On average, a reduction of 25-200 bp with each cell division *in vitro* and between 30-60 bp per year *in vivo* is observed. Healthy young individuals show a mean telomere length (TL) of 10000 bp compare to elderly who have lengths between 5000-7000 bp [5]. Different cell types have also shown variations in the rate of telomere reduction, being in peripheral blood leukocytes 20-40 bp/year [6-8], skin fibroblasts 19 bp/year [9], adult bone marrow 9 bp/year [10], all of them associated to their replicative potential.

Human cells have at least two mechanisms to maintain their telomeres: one more common mediated by telomerase, and another one less frequent, called ALT (Alternative Lengthening of Telomeres), based on homologous recombination [11-13]. The enzyme telomerase is a ribonucleoprotein complex that compensates the telomere reduction by adding new repeats to chromosome ends. Telomerase is composed of two subunits: the human telomerase reverse transcriptase (hTERT), which has the catalytic activity and the RNA component (*hTERC*) that provides the template for the telomeric synthesis. Other proteins like dyskerin, NOP10 and NHP2 associate with telomerase complex and are particularly important for TERC stabilization and also play an important role in ribosome biogenesis and RNA processing. Since telomerase activity is absent in most normal somatic cells, but it is present in most of malignant tissues and immortal human cell lines, activation of telomerase may be therefore a critical step in human cancer development [14, 15]. Moreover, telomerase is reactivated in over 90% of human tumors and in a large percentage of immortalized cell lines [16, 17]. In telomerase-positive cells, telomeres are usually short and homogeneous in size. In reference to ALT, it has been found to maintain telomeres in at least 30-40% of immortalized human cell lines and 10-15% of human tumors [12, 18]. A number of recombination proteins have been shown by genetic analyses to be necessary for telomere maintenance in ALT cells, however, the precise mechanism is not yet well known.

Telomeric DNA repeats are also maintained by six telomere-associated proteins -TRF1, TRF2, POT1, RAP1, TIN2 and TPP1 in mammalian cells- called the shelterin complex [19, 20]. It is essential for telomere function and enables cells to distinguish their natural chromosome ends from breaks in the DNA by repressing the DNA damage response at telomeres, as well as regulating telomere length [21]. Particularly, Telomere repeat factor-1 (TRF1) and -2 (TRF2) are related proteins that share sequence similarities, and along with Protection of Telomeres-1 (POT1), they interact directly with telomeric DNA. Regarding their functions, TRF1 is involved in a negative feedback mechanism that allows telomere shortening by inhibiting the telomerase activity [22] whereas TRF2 participates in the t-loop formation, capping and protecting the 3' single strand overhang. It has been shown that the amount of telomere-bound TRF1 correlates with telomere length. In addition, an increased expression of both *TRF1* and *TRF2* shortens telomeres, but loss of its activity leads to telomere-telomere fusion events, suggesting a protective role for both proteins in the maintenance of telomere structure and function [20]. As for TIN2 (TRF1-interacting nuclear factor 2), it binds TRF1 and TRF2 through independent domains and recruits the TPP1 (TIN2- and POT1-interacting protein 1)-POT1 (protection of telomeres 1) complex, constituting the bridge among the different shelterin components [23-25]. In addition, the N-terminus of TPP1 contains a telomerase-interacting domain, suggesting a role for TPP1 in the recruitment of telomerase to chromosome ends [26]. Finally, RAP1 (repressor activator protein 1) forms a complex with TRF2 and this association is essential for RAP1 binding to telomeres [27-29].

Besides the shelterin complex, mammalian telomeres interact with a number of other factors that can influence chromosome integrity, such as tankyrase 1 and tankyrase 2, meiotic recombination 11 homologue (MRE11), the RecQ-like helicases WRN (Werner's syndrome protein) and BLM (Bloom's syndrome protein), Ku70, Ku86, ataxia-telangiectasia mutated (ATM) and the DNA-repair protein RAD51D7 [30]. These factors are involved in DNA recombination and repair, suggesting that these machineries are an important component of telomere replication, protection and stability. Among them, TANK1 (tankyrase-like protein 1) is a member of the growing family of poly (ADP-ribose) polymerases (PARPs) that interacts with and ADP-ribosylates the telomere-binding protein TRF1 [31]. *In vitro*, ribosylation by TANK1 displaces TRF1 from telomeric DNA, suggesting that TANK1 might be a positive regulator of TL in telomerase-expressing cells [22]. There are also studies that confirm this role *in vivo* [32]. Tankyrase over-expression in human cells induces a progressive elongation of telomeres and is expected to be upregulated in all human tumors.

Telomere dysfunction, including shortening in length or structural changes, has been shown to contribute to cancer. Different studies have demonstrated that the decline of telomere length contribute directly to the presence of chromosomal abnormalities usually found in several types of cancer. Moreover, a number of factors that increase the probability of telomeric associations, like critical short telomeres or mutations in telomere associated genes, have been identified [33, 34]. These telomeric associations are the basis of genomic instability as they lead to anaphase bridging and subsequent breakage, which requires further repair. Repetition of this breakage-fusion-bridge cycle causes aneuploidy and chromosome recombination leading to a reorganization of the cell genome that could result in tumorigenesis [35-37].

As mentioned above, normal human cells express low to undetectable levels of telomerase activity, whereas more than 90% of human tumor cells express high levels of this enzyme [17, 38]. Because tumor cells typically show shorter TL with respect to neighboring normal cells, telomerase activation is thought to be a relatively late event in the transformation process that may stabilize telomere lengths [37]. In addition, telomere reduction and its consequent genomic instability are thought to promote the transformation process. Different studies support this, by using telomerase knockout mice, which display progressive telomere shortening, similar to the behavior of telomeres in normal cells [39]. These animals do not exhibit any apparent pathological phenotypes in the early generations, in contrast with late-generation mice that display pathologies associated with aging and demonstrate an enhanced transformation phenotype. They develop lymphomas and teratocarcinomas with high frequency, associated with extensive anaphase bridge formation and genetic instability [37]. Mice deficient in both TERC and p53 develop an even more varied tumor spectrum, displaying lymphomas, sarcomas, and adenocarcinomas [40].

Lymphoid Malignancies

Lymphoid malignancies constitute a heterogeneous group of lymphoproliferative disorders that represent distinct disease entities, according to their clinical, morphological, immunophenotypic and genetic features [41]. Among them, our interest was focalized in the study of some subtypes of non-Hodgkin lymphomas, plasma cell disorders and chronic lymphocytic leukemia.

Non-Hodgkin Lymphomas (NHL)

NHLs comprise a group of closely related heterogeneous diseases derived from B or T/NK cells. B-cells NHL represent the most common malignant lymphoid neoplasms, with the majority of them arising from germinal centre or post-germinal centre B cells, due to a disruption of the different phases of normal B-cell development. They show variable clinical behavior ranging from highly aggressive to an indolent course [41]. Data from the literature showed an increased incidence of NHL in the world, particularly the most aggressive histological subtypes [42].

Follicular Lymphoma (FL)

FL is the most prevalent form of low grade B-cell lymphoma in adults, accounting for 20-35% of newly diagnosed adult NHL in Western countries, and representing the second most frequent NHL subtype [41]. It is a neoplasm of follicle centre B-cells. Clinically most FL patients are elderly and present with advanced stage disease. Median survivals approximate 8–10 years, as the disease runs an inexorable course in most cases. The majority of patients (25%-60%) are at risk for histologic progression to a process resembling diffuse large B cell lymphoma, associated with a rapidly progressive clinical course, refractoriness to treatment and short survival. The broad spectrum of clinical behaviour in FL, together with the slope of the survival curve, provides compelling evidence for the marked clinical heterogeneity of this tumor. FL is genotypically associated to translocation t(14;18)(q32;q21) (70-95% of cases) that determine the juxtaposition of *BCL2* protoncogene in 18q21 with the immunoglobulin heavy chain (*IGH@*) gene *locus* at 14q32. On chromosome 18, the breakpoints occur at two *loci*, the major breakpoint region (MBR) located in the 3´-untranslated region of the third exon of the *BCL2* gene (60-80% of cases), and the minor cluster region (mcr), located in a DNA stretch, approximately, 30 kb downstream of the BCL2 locus (10-15% of cases). On chromosome 14, breakpoints are predominantly found within the joining region (JH) of the *IGH@* gene [43, 44]. Translocation t(14;18) results in deregulated expression of the *BCL2* gene, which encodes a 26 kDa protein that extends cell survival by blocking programmed cell death. BCL2 also has a role in DNA damage, avoiding DNA repair, and leading to increased genomic instability [45, 46].

Diffuse Large B-Cell Lymphoma (DLBCL)

DLBCL is the most prevalent form of NHL in Western countries, representing 40% of all adult cases. This group of tumors encompasses heterogeneous clinical, morphological, immunophenotypic, cytogenetic and molecular genetic features that are reflected in a highly variable clinical course. DLBCL may arise from *de novo* or, alternatively, may develop from the transformation of a FL [41]. DLBCL remains an important clinical challenge since at least one third of patients are not cured by currently available therapeutic regimens, including combined immune-chemotherapy. Approximately, 50% of cases exhibit chromosomal translocations including one of the immunoglobulin gene sites, which lead to deregulated expression of a variety of genes [47, 48]. Studies employing DNA microarrays identified two major molecular categories with gene expression patterns suggestive of different stages of B-cell development. One type had an expression profile characteristic of germinal centre B-cells, associated to a better outcome, whereas the other type had a profile similar to that of *in vitro* activated peripheral blood B-cells, and is related to poor clinical evolution [49, 50]. More recently, gene expression profiling identified primary mediastinal B-cell lymphoma as a clinically favorable subgroup of DLBCL [51].

Mantle Cell Lymphoma (MCL)

MCL is a well defined lymphoid neoplasm that accounts for approximately 5 to 10% of all NHL. It is characterized by a male predominance, frequently advanced clinical stage at diagnosis, and a poor prognosis with a median survival range of 3-4 years [41]. MCL shows architectural destruction by a monomorphic lymphoid proliferation with a nodular, diffuse or mantle zone growth pattern. Blastoid variants both classic and pleomorphic are considered to be of potential clinical significance. The genetic hallmark is the translocation t(11;14)(q13;q32) that juxtaposes the proto-oncogen *CCND1* at chromosome 11q13, to the *IGH@* gene at chromosome 14q32, leading to the constitutive overexpression of cyclin D1, a cell cycle regulator that is normally not expressed in B-cells. Most of MCL showed no mutations in variable sequences of immunoglobulin (*IGVH*) gene according to a pregerminal-center cell origin. However, up to one-third of cases display mutated *IGVH* genes suggesting that some tumors may originate in specific subsets of B cells that have entered the follicular germinal center [52]. No association between *IGVH* mutational status and outcome was observed [53].

Plasma Cell Disorders

Include a wide range of pathologies ranging from *monoclonal gammopathy of undetermined significance* (MGUS) to *multiple myeloma* (MM). MM constitutes approximately 10% to 15% of all hematologic malignancies and about 1% of all forms of cancer. It is characterized by atypical plasma cell infiltration into the bone marrow (BM) and the presence of monoclonal protein in serum and/or urine. It arises from a stepwise neoplastic transformation process with accumulation of progressive genetic events that favor proliferative advantages and the expansion of the mutant cell clone [54]. Genomic instability, including translocations at *IGH@* gene and overexpression of cyclin D are considered pathogenetic in this disease [55, 56]. The clinical manifestations counting osteolytic lesions, anemia, immunodeficiency, and renal abnormalities can be attributed to underlying plasma cell proliferation [55]. The outcome for patients with MM is highly variable. Although the median overall survival time is 3 to 4 years, the range is from less than 6 months to greater than 10 years. This variability derives from heterogeneity in both myeloma cell biology and multiple host factors.

The natural course of the disease may progress from MGUS, a pre-symptomatic phase, to MM. MGUS is characterized by a monoclonal protein in serum or urine without other clinical features of MM. It is one of the most common premalignant disorders in western countries with a prevalence of 3.2% in the Caucasian Population 50 years of age or older. The transformation rate of MGUS to MM is about 1% per year with an actuarial probability of malignant evolution of 30% at 25 years. After a median of 10 years about a quarter of MGUS patients develop MM [57].

Chronic Lymphocytic Leukemia (CLL)

CLL is the most frequent type of adult leukemia in Western world. The disease is characterized by the accumulation of long-lived circulating clonal leukemic B-cells resulting from a complex balance between cell proliferation and apoptotic death. CLL is rare in people younger than 50 years, showing a mean age of 65-70 years according to different studies. CLL patients show a highly variable clinical course, with time to progression ranging from months to decades.

While some patients require therapy immediately after diagnosis, others have a more stable, indolent disease without need treatment for decades [58]. Analysis of genomic abnormalities using interphase fluorescence *in situ* hybridization (FISH) has resulted in detection of clonal aberrations in more than 80% of CLL patients making possible to establish high-risk groups of prognostic value [59].

Monoallelic deletion of 13q14.3 is the most frequent abnormality, occurring in approximately 50% of CLL patients, and is associated with favorable clinical course when presented as a sole alteration. Deletions of 11q22 (involving *ATM* gene) and 17p13 (including *TP53* gene) have the shortest median survival, while trisomy 12 and biallelic 13q14.3 deletion correlate with intermediate prognosis [60, 61]. Translocations are rare events in CLL. They are usually unbalanced resulting in loss of genetic material, and have been associated to poor outcome [62, 63].

Telomeres in Lymphoid Malignancies

Several studies have shown that short telomeres and high telomerase activity are associated with tumorigenesis and poor prognosis in patients with hematological neoplasm [14, 5]. More precisely in lymphoid malignancies the mean TL was reported as an emerging prognostic factor in patients with CLL, in which cells of the progressive form of this disease have shorter telomeres than those of the indolent form [64, 65]. In addition, short telomeres may be associated with genetic complexity and a high risk of genomic aberrations in this entity [66].

In B-cell NHL, Ladetto et al. [67] have found a correlation between TL and tumor histopathogenesis in relation to the germinal center (GC) origin, whereas other authors including our group [68, 69] suggested that TL is a more dynamic factor that also would reflect the telomere shortening and lengthening process. In this regard, a positive correlation of telomerase activity with the rate of proliferation in B-cell NHL was found [70-73]. In plasma cell disorders, a few reports have evaluated TL and telomerase activity in patients with MM [74-76], showing a negative correlation between both parameters. In reference to telomere-associated proteins, Poncet et al. [77] found low expression of many components involved in telomere elongation and capping, as well as an increase in telomerase activity in CLL cells, suggesting a global deregulation associated to the pathogenesis of the disease. Moreover, in NHL high expression of *TRF2* and *hPif* were associated to high telomerase activity, supporting that stabilization of shorter tumor telomeres might need higher levels of telomere-binding protein expression [78].

In this study, a comparative analysis of telomere length in different lymphoid malignancies was performed and results were correlated with genetic abnormalities and gene expression profiles of telomere-associated proteins. A comparison with the literature was also performed.

Materials and Methods

Patients

The present study included 262 patients with different B-cell lymphoid malignancies. Diagnoses, number of patients, distribution by sex and age and type of sample are summarized in Table 1. All individuals provided their informed consent according to institutional guidelines. The study was approved by the Ethics Committee of our Institution.

Table 1. Characteristics of patients and samples

Group	No. of cases	Mean age (years) (range)	Sex		Sample, No.		
			F	M	BM	LN	PB
NC	30	65.0 (33-86)	17	13	-	-	30
FL	44	54.2 (29-78)	22	22	33	11	-
DLBCL-S	9	62.5 (53-75)	3	6	6	3	-
dn-DLBCL	11	51.9 (22-69)	5	6	7	4	-
MCL	20	60.0 (31-82)	5	15	14	6	-
CLL	13	68.0 (55-87)	9	4	-	-	13
MM	99	66.2 (30-87)	52	47	99	-	-
MGUS	66	69.3 (39-88)	37	29	66	-	-

NC: normal controls; FL: follicular lymphoma; DLBCL-S: diffuse large B-cell lymphoma secondary to FL; dn-DLBCL: *de novo* DLBCL; MCL: mantle cell lymphoma; CLL: chronic lymphocytic leukemia; MM: multiple myeloma; MGUS: monoclonal gammopathy of undetermined significance; F: female; M: male; BM: bone marrow; LN: lymph node; PB: peripheral blood.

Telomere Length Evaluation

Telomere length was evaluated by terminal restriction fragments (TRF) assay as previously described [75]. Briefly, high-molecular weight DNA was extracted from BM, lymph node (LN) and peripheral blood (PB) samples. Aliquots of 10 μg DNA were digested with *HinfI* and *RsaI* restriction enzymes at 37ºC, overnight. The digested products were separated by electrophoresis in 0.8% agarose gels during 20h at 35V. Gels were first depurinated, then denatured and neutralized and finally transferred to nylon membranes by Southern blotting. The telomeric fragments were detected through hybridization with a 5' end-labeled telomeric probe $(TTAGGG)_7$. The hybridized membranes were exposed to films during 10-15 days at -70ºC before developing. Hybridization signals were evaluated in the autoradiograph by densitometric scanning in each lane respect to a λ/*HindIII* molecular weight standard and were analyzed using the Gel Pro software (Media Cybernetics). Telomere mean values were measured by estimating the band size corresponding to the point with the highest intensity. For comparative analysis, peripheral blood mononuclear cells from 30 healthy individuals (13 males and 17 females; median age 65 years, range: 33-86 years) as normal controls were also evaluated. In addition, as internal controls for telomere shortening we used K562 cell line (positive control of telomere shortening), and cord blood cells (negative control) as a very young tissue without telomere reduction.

RNA Extraction, Reverse Transcription and Real-Time Quantitative RT-PCR

Total RNA was extracted by conventional methods from mononuclear cells isolated from BM samples of patients and PB of normal controls, and from K562 cell line. RT-PCR was carried out using 1X RT Buffer, 200 U/μl of Moloney Murine Leukemia Virus Reverse Transcriptase, 250ng/μl random primer and 10 mM each dNTP. The cDNA synthesis was performed in a total volume of 20 μl, containing 1μg of the total RNA, for 10 minutes at 95ºC, for 60 minutes at 37ºC and 10 minutes at 95ºC to inactivate the enzyme. cDNA was stored at -20ºC until use.

The analysis of *TRF1*, *TRF2*, *TANK1* and *hTERT* mRNAs were performed using real-time quantitative PCR in a LightCycler system (Roche Diagnostics), based on TaqMan methodology. The mRNA expression levels were measured using gene-specific fluorescent-labeled probes. FAM was added as the 5'-fluorescent reporter while TAMRA was added to the 3' end as a quencher. Primers and probes sequences were previously described [78, 79]. The housekeeping gene *GAPDH* (glyceraldehyde 3´phosphate dehidrogenase) was used to normalize sample-to-sample differences in cDNA input, RNA quality and RT efficiency, and it was amplified using the primers and probe described by Hu et al [79]. The PCR reaction was done using 4 μl of each RT reaction, 1X TaqMan master mix (Roche Diagnostics), 200 nmol/L of the probe and 300 nmol/L of TRF1 primers and 500 nmol/L of TRF2, TANK1, hTERT and GAPDH primers in a 20 μl final volume. For all targets, the thermal cycling conditions were 10 minutes at 95ºC to activate the enzyme, followed by 45 cycles at 95ºC for 15 seconds and 60ºC for 1 minute. All measurements included a determination of the standards and no-template as a negative control, in which water was substituted for the cDNA. Standard curves were constructed with fivefold serial dilutions of the cDNA from the K562 cell line (5 μg to 0.15 μg total RNA). The ratio between copy numbers of *GAPDH* and *TRF1*, *TRF2*, *TANK1* or *hTERT* genes in samples and controls were calculated with the LightCycler software, using these standard curves.

Cytogenetic Analysis

Chromosome analyses were performed on PB lymphocytes for CLL patients and BM cells for MM and MCL cases. PB lymphocytes were cultured for 96 hours with mitogen stimulation and BM cells by direct method or short term culture (24-48 hs), both at 37ºC in F-12 medium supplemented with 15% of fetal calf serum. Slides were prepared by conventional method. G-banding technique was used. Karyotypic abnormalities were described using the International System for Human Cytogenetic Nomenclature [80].

DNA Extraction and Nested PCR for BCL2/IGH@ Rearrangement

Genomic DNA was isolated from fresh tissues and peripheral blood lymphocytes by conventional Phenol-Chloroform extraction and ethanol precipitation. DNA concentration and purity were estimated by agarose gel electrophoresis and UV spectrophotometry. The quality of all DNA samples was further checked by PCR amplification of a target DNA within a β-globin gene fragment (536 bp). PCR detection for the MBR-JH and mcr-JH rearrangements of the t(14;18) translocation were performed as previously described [81].

Statistical Evaluation

The comparison of data from patients and controls and between subgroups was performed using Mann-Whitney test or the Student t test. Correlations between gene expression and TL were assessed by using the Kendall's coefficient. The cut-off point for *hTERT* expression was selected according to receiver operating characteristic (ROC) analysis. For all tests, $p < 0.05$ was regarded as statistically significant.

RESULTS

TRF Lengths in B-Cell Malignancies

From a total of 262 patients, TL was evaluated in 220 cases with different B-cell malignancies: 20 MCL, 44 FL, 9 DLBCL secondary to FL (DLBCL-S), 11 *de novo* DLBCL (dn-DLBCL), 13 CLL, 78 MM and 45 MGUS. Telomere lengths from K562 cell line and cord blood were 5.2±0.1 kb and 14.5±1.1 kb, respectively.

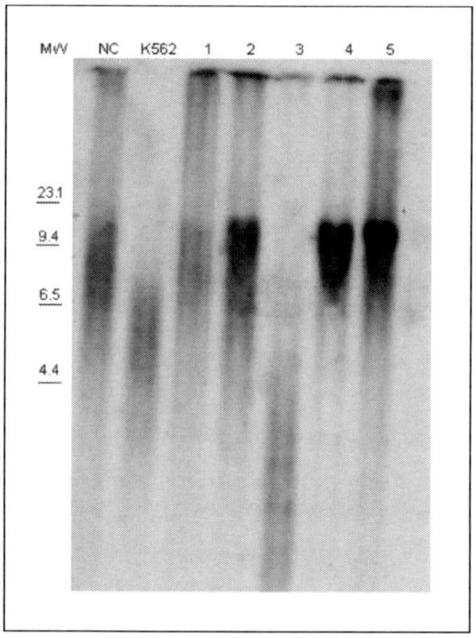

Figure 1. Representative Southern blot analysis in FL patients and controls. MW: molecular weight; NC: normal control; Lanes 1-5: FL patients.

In none of pathologies, correlations with age and sex were found. All of them showed significant differences compared to normal controls (8.12+0.26 kb) (p=0.03). A representative example of TRF length evaluation is shown in Figure 1.

The analysis performed separately in each disease showed significant differences in mean telomere length among them (Table 2, Figure 2). Patients with DLBCL-S had the shortest telomeres, followed by FL [82]. MCL [69] and dn-DLBCL showed similar mean TL, whereas MGUS [83] and CLL cases displayed the longest TL. Mean TRF values for MM were intermediate among those entities. Moreover, DLBCL-S had significant differences with respect to all other entities: FL (p=0.01), dn-DLBCL (p=0.019), MCL (p=0.0001), CLL (p=0.0001), MM (p=0.007) and MGUS (p=0.003). TRF length in FL was also significantly different from that in CLL, MM and MGUS (p=0.0001). Finally, differences in dn-DLBCL and MCL with respect to CLL (p=0.021 and p=0.0001, respectively) and MGUS (p=0.04 and p=0.0004, respectively) were observed, as well as between MM and MGUS (p=0.027).

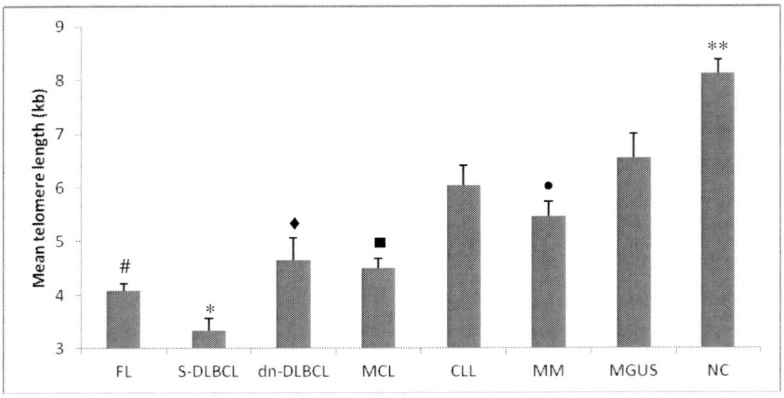

Figure 2. Mean telomere length according to the subtype of the lymphoproliferative disorders. ** Significant differences respect to NC: p=0.03; * to all the other subtypes: p<0.01; # to CLL, MM and MGUS: p<0.0001; ♦ to CLL and MGUS: p<0.04; ■ to CLL and MGUS: p<0.004; ● to MGUS: p=0.027.

Table 2. Mean TRF values and ranges for each entity

Group (n)	TRF (kb)	Range (kb)
NC (30)	8.12+0.26	(7.00-11.21)
FL (44)	4.09±0.12#	(2.47-6.30)
DLBCL-S (9)	3.34±0.22*	(2.47-4.64)
dn-DLBCL (11)	4.65±0.42♦	(3.05-6.42)
MCL (20)	4.51±0.17■	(3.57-5.90)
CLL (13)	6.04±0.37	(4.28-9.15)
MM (78)	5.47±0.26●	(2.69-9.44)
MGUS (45)	6.55±0.45	(3.28-14.6)

NC: normal controls; FL: follicular lymphoma; DLBCL-S: diffuse large B-cell lymphoma secondary to FL; dn-DLBCL: *de novo* DLBCL; MCL: mantle cell lymphoma; CLL: chronic lymphocytic leukemia; MM: multiple myeloma; MGUS: monoclonal gammopathy of undetermined significance.*Significant differences respect to all the other subtypes: p<0.01; # to CLL, MM and MGUS: p<0.0001; ♦ to CLL and MGUS: p<0.04; ■ to CLL and MGUS: p<0.004; ● to MGUS: p=0.027.

Telomere Length and Genetic Characteristics

The correlation with cytogenetic characteristics was evaluated in patients with MM, MCL and CLL. For MM cases, cytogenetic studies were performed in 30 cases. Nine of them (30%) had clonal chromosome abnormalities, showing a TRF mean value significantly shorter than that observed for MM patients with normal karyotypes (p=0.04). Likewise, in

CLL significantly shorter TL in patients with chromosome abnormalities compared to cases with normal karyotype was observed (p<0.03) (Table 3). In addition, for MCL we compared TL of patients with (4.84±1.09 kb) and without (4.40±0.68 kb) complex karyotypes, and no significant differences between them were observed (p=0.43) [69].

In FL and DLBCL-S, TRF data were also correlated with the molecular evaluation of *BCL2* gene rearrangements [82]. Eighty four percent of FL patients showed *BCL2* gene rearrangements and seven cases (15.9%) were *BCL2-IGH@* negatives. Patients positive for *BCL2* gene rearrangements showed larger TRF values than did *BCL2*-negative patients (p<0.05). In DLBCL-S, 77.7% (7/9) cases were *BCL2*-positive. No differences between TRF values in cases with and without t(14;18) in this small sample of DLBCL-S patients were found (Table 4).

In FL, TRF values were also analyzed according to the breakpoint sites involved [82]. MBR-JH was observed in 72.3% of patients, meanwhile 11.3% showed mcr-JH rearrangement. We found shorter telomere lengths in *BCL2*-negative patients (3.42±0.14 kb), intermediate values when breakpoints occurred in mcr-positive cases (3.97±0.33 kb), and larger values when breakpoints occurred in MBR-positive patients (4.24±0.15 kb), with significant differences between MBR-positive and *BCL2*-negative cases (p=0.033). Note that TRF values from *BCL2*-negative patients were similar to those observed in DLBCL-S.

Table 3. Mean TL according to cytogenetics

Karyotype	Mean telomere length (kb)	
	MM	CLL
Normal	6.29±0.44	6.93±0.40
Abnormal	4.76±0.57*	5.22±0.51#

MM: multiple myeloma; CLL: chronic lymphocytic leukemia. Significant differences respect to normal karyotype: *p=0.04; #p<0.03.

Table 4. Analysis of mean TL and *BCL2* rearrangement

NHL	*BCL2* gene rearrangement	
	positive	negative
FL	4.19±0.14*	3.42±0.14
DLBCL-S	3.63±0.85	3.08±0.17

FL: Follicular lymphoma; DLBCL-S: diffuse large B cell lymphoma secondary to FL. Significant difference respect to *BCL2* negative FL *p<0.05.

Telomeric Protein Expression

The mRNA expression of *TRF1, TRF2, TANK1* and *hTERT* was evaluated in 64 patients with MGUS and 68 with MM [83]. A differential expression pattern with lower levels for *TRF1* and higher for *TRF2, TANK1* and *hTERT* in patients with respect to controls were observed (Figure 3).

When both pathologies were compared, a significant increase of *TRF1* mRNA levels (p=0.006) and a lower expression of *TRF2* (p=0.005) and *TANK1* (p=0.003) in MGUS with respect to MM patients were found. No difference in *hTERT* values between MGUS and MM was observed.

In both entities, the expression of *hTERT* was widespread, with patients that over-expressed this gene and others that showed undetectable levels. Thus, for a better analysis patients were divided into three groups: low (≤1.08) (group I; GI), intermediate (<1.08 to ≥5.0) (group II; GII) and high (>5.0) (group III; GIII) mRNA expression (Table 5).

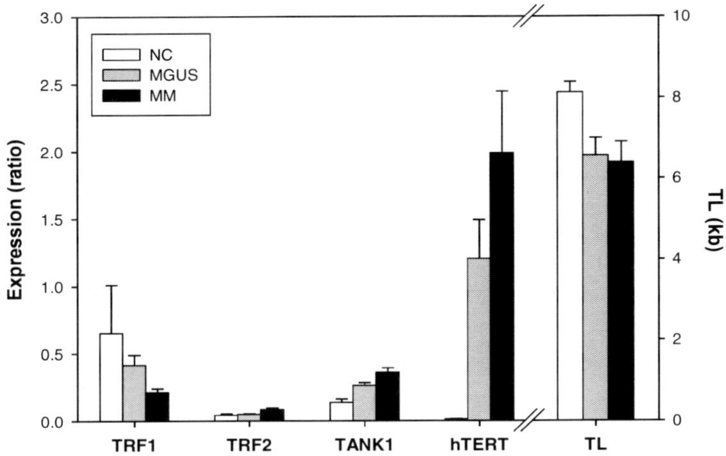

Figure 3. Histogram showing means TL and mRNA expression profiles of *TRF1*, *TRF2*, *TANK1* and *hTERT* genes in patients and normal controls (NC). Significant differences with respect to NC: *p<0.01 and to MM: #p<0.006 [83].

Both entities demonstrated a similar distribution of patients per group, with most of the cases in GI (50 MGUS and 44 MM) and the least in GIII (5 MGUS and 7 MM). The analysis of TL according to *hTERT* groups showed that GIII, with the highest telomerase expression, had the shortest TL with significant differences respect to GII (MGUS: p=0.03; MM: p=0.04) and GI (MM: p=0.04). Moreover, in MM the GIII mean TRF value was also significantly reduced compared to K562 cell line (5.02±0.3kb) (p=0.04) [83].

The evaluation of remaining genes according to *hTERT* groups showed in both pathologies increased levels of *TRF2* and *TANK1* from GI to GIII. Significant differences for both genes in MM (GIII versus GI and GII: *p* <0.01) and for *TRF2* in MGUS (GIII versus GI: p<0.01) were observed.

In addition, the comparison between entities in GIII showed that both genes had lower mRNA expression in MGUS respect to MM, with significant differences for *TRF2* (p=0.01). Regarding *TRF1*, a non-significant increase from GI to GIII in MM was observed.

Also, we explored the relationship among mRNA expression of different genes. Both entities showed a positive association between *TRF2-TANK1* (p<0.0001), *TRF2-hTERT* (p<0.0001) and *TANK1-hTERT* genes (p<0.0001 for MM and *p*=0.01 for MGUS) [83].

Table 5. Analysis of telomeric genes expression and mean TL according to groups of *hTERT* expression in patients with MM and MGUS [83]

Groups	Telomeric Genes (X±ES)				TL (kb) (X±ES)
	hTERT	*TRF1*	*TRF2*	*TANK1*	
Multiple myeloma					
GI	0.30±0.04	0.18±0.04	0.06±0.01	0.28±0.07	7.04±0.51
GII	2.17±0.24	0.21±0.03	0.07±0.01	0.39±0.05	6.76±0.79
GIII	12.10±1.52	0.34±0.08	0.22±0.03#	0.79±0.16#	4.14±0.63*
Monoclonal gammopathy of undetermined significance					
GI	0.27±0.03	0.38±0.08	0.04±0.004	0.24±0.01	7.17±0.41
GII	2.53±0.28	0.65±0.28	0.05±0.02	0.26±0.08	6.92±0.31
GIII	8.05±1.28	0.23±0.06	0.10±0.02#°	0.39±0.11	4.93±1.01**

-Significant differences of GIII respect to GI and GII in MM and respect to GI in MGUS: #p<0.01.-Significant differences of GIII between MGUS and MM: °p=0.01.-Significant differences of GIII respect to GI and GII: *p=0.04, and respect to GII: **p=0.03.

Comparison with the Literature

Our results of telomere shortening in different B-cell malignancies were compared with data from the literature [67, 68] (Table 6).

Table 6. Telomere length in different series of patients with B-cell malignancies

Subtype	Ladetto et al. [67]		Walsh et al. [68]		Our group [69, 75, 82 and the present study]	
	N° of cases	TRF (kb)	N° of cases	TRF (kb)	N° of cases	TRF (kb)
CLL	34	4.35	36	4.50	13	6.04
MCL	10	3.58	57	4.52	20	4.51
FL	33	7.38	18	4.71	44	4.09
DLBCL-S	-	-	-	-	9	3.34
dn-DLBCL	14	7.78	93	4.71	11	4.65
BL	5	9.47	-	-	-	-
MZL	9	5.96	-	-	-	-
MM	18	6.28	-	-	78	5.47
MGUS	-	-	-	-	45	6.55
HCL	-	-	12	5.03	-	-
LPL/WM	-	-	7	6.38	-	-

CLL: chronic lymphocytic leukemia; MCL: mantle cell lymphoma; FL: Follicular lymphoma; dn-DLBCL: *de novo* diffuse large B cell lymphoma; DLBCL-S: DLBCL secondary to FL; BL: Burkitt lymphoma; MZL: marginal zone lymphoma; MM: multiple myeloma; MGUS: monoclonal gammopathy of undetermined significance; HCL: hairy cell leukemia; LPL/WM: lymphoplasmacytic lymphoma/Waldenström´s macroglobulinemia; TRF: terminal restriction fragments.

Ladetto et al. [67] found the shortest TRFs in MCL and CLL, meanwhile FL, DLBCL and BL, had the largest telomeres, and MZL and MM showed intermediate TL. Walsh et al. [68] found that CLL and MCL had the shortest TRFs, DLBCL and FL showed somewhat largest telomeres, whereas hairy cell leukemia and lymphoplasmacytic lymphoma (HCL)/Waldenström´s macroglobulinemia (WM) displayed the longest telomeres. Our results showed the shortest telomeres in DLBCL-S, a subtype not studied by other groups. TRF data from MCL, DLBCL and FL were comparable in our group and those of the study of Walsh et al. [68].

On the contrary, results from MM were lower than those observed by Ladetto et al. [67] and TRF from CLL were larger in our group compare to the other studies. MGUS was only analyzed in our series and showed the largest TL.

Discussion

In the present work, we analyzed the telomere erosion in different B-cell malignancies and its association with genetic characteristics and telomere-associated proteins expression. Telomere length was found to be highly heterogeneous and significantly reduced in B-cell disorders compared to controls.

A number of reports have investigated the telomere dysfunction in a large panel of different lymphoid malignancies [67, 68]. Ladetto et al. [67] found a correlation between telomere length and tumor histopathogenesis in relation to the GC origin. Thus, the shortest TRFs were observed in pre-germinal tumors, like MCL and CLL, GC-derived neoplasms showed the largest telomeres, and post-GC-derived malignancies had intermediate TL. This correlation was not observed either by Walsh et al. [68] or by our group. Walsh et al. [68] found similar TRF values in CLL, MCL, FL and DLBCL, and largest TL in HCL and WM, and they suggested that TL is a more dynamic factor that do not correlate simply with GC-related origin but also would reflect the telomere shortening and lengthening process. In our experience [69, 75, 82, and the present study], MGUS and CLL had the largest telomeres, followed by MM, dn-DLBCL, MCL and FL. The shortest TRF was observed in DLBCL-S, a very aggressive entity. Largest TRF length in our CLL patients may be probable related to the reduced number of cases analyzed and/or their clinical characteristics. Our results in MCL and DLBCL were similar to those observed by Walsh et al. [68], but differences for FL and CLL were found. Studies of Ladetto et al. [67] and Walsh et al. [68] showed comparable results only for CLL. This high variability reflects the heterogeneous characteristics of patients with B-cell disorders and probably, the different proliferative activity in tumors during malignant transformation.

As known, telomere shortening is associated with genomic instability, evidenced by increased frequencies of telomeric associations, contributing to the presence of chromosome abnormalities usually observed in diverse types of cancer [84]. In our study, we correlated TL with cytogenetic characteristics in patients with MM, CLL and MCL. In MM, the mean TRF values of patients with clonal chromosome abnormalities were significantly lower than that observed for those with normal karyotypes. These findings agree with results observed by Wu et al. [76] who found a negative correlation between cytogenetics abnormalities and the coexistence of high telomerase activity and short TL. As known, MM patients with abnormal karyotypes have a worse prognosis than those with a diploid pattern [85, 86]. Thus, the association of abnormal karyotypes with a remarkable telomere reduction supports the importance of this mechanism in the development and/or progression of the disease.

With regards to CLL, although the number of cases from our series is too small yet, significant differences in TL between patients with and without chromosomal abnormalities were observed. There are studies correlating telomere length and genomic aberrations detected by FISH analysis [65, 66] but, to our knowledge, the association between abnormal karyotypes and telomere length in CLL has not been evaluated. In line with our findings, Lin et al. [87] have demonstrated that short dysfunctional telomeres, complete telomere loss, and telomere fusion are common events in CLL patients and become more common in advanced disease. In addition, we compared TRF in MCL patients with and without complex karyotypes, and no significant difference between them was found. These findings would reflect the disease homogeneity and the high genomic instability of this aggressive disease, and are in line with the comparable TRF values detected by Walsh et al. [68] for MCL cases with and without mutations in *IGVH* genes. More studies are necessary to confirm these results.

In reference to FL subtype, we have reported [82] significant differences in TL between *BCL2* positive and negative FL patients and also an association between TRF values and *BCL2* breakpoints. There are few reports comparing cytogenetic and/or clinical features of FL patients with and without t(14;18). Different patterns of recurring chromosome alterations associated with FL with and without t(14;18) has been identified [88-90], indicating the deregulation of different genetic pathways in each lymphoma subset and suggesting the existence of distinct subtypes of lymphomas originated from the germinal centre. Simultaneously, the clinical signification of the lack of t(14;18) in FL is a controversial point with some authors finding different clinical characteristics and others not [90-93]. Respect to the correlation of TL with the breakpoints involved in the *BCL*2 rearrangements we have observed shorter telomeres in cases without the translocation, intermediate in mcr positive patients and longer in MBR positive cases. Up to date, there are not a clear relation between *BCL*2 breakpoints and clinical characteristics. Interestingly, DLBCL-S showed the shortest TL, suggesting the participation of telomere shortening in the transformation process from FL to a DLBCL, entity associated with a rapidly progressive clinical course, refractoriness to treatment and short survival.

In reference to MGUS, there is very scarce information in the literature about TL, being our study the first one with an important number of evaluated patients [83]. It showed the longest mean TL among all the entities analyzed, with significant differences respect to normal controls, consistent with its condition of pre-malignant disease. In addition, the significantly shorter TL in MM with respect to MGUS might suggest the participation of telomere reduction in the transformation process from MGUS to MM.

As known, during tumorigenesis, telomere short progressively up to the telomerase activity is restored, allowing the cells to proliferate indefinitely. That activation is responsible of the presence of short telomeres in neoplasias in contraposition to the high telomerase activity. This peculiarity of tumor cells suggests that telomerase inhibition may represent an effective therapeutic approach to diminish their replication and inducing a selective death [94, 95]. The significance of telomerase expression in human cancer has been evaluated in different lymphoid malignancies [70, 74, 76]. Besides, high levels of telomerase activity were also found in benign lymphoid tissues, leading to the suggestion that telomerase expression in malignant B cells can be explained by an induction-and-retention model of reactivation of telomerase in malignant clones [70]. In B-cell NHL, an increased telomerase activity compared to reactive lymph nodes was reported. Differences of telomerase expression between the high- and low-grade lymphomas were also observed, reflecting the varying demands of telomerase activity to prevent telomere shortening [70]. Furthermore, a positive correlation of telomerase activity with the rate of proliferation in the B-cell NHL was found [71-73]. In this context, Klapper et al. [78] detected an up-regulation of telomerase in Burkitt's lymphoma with respect to the activity observed in reactive lymph nodes, MCL, FL and DLBCL.

In addition, elevated telomerase levels in CLL and HCL were reported compared to normal B lymphocytes [96, 97]. Short telomeres and high telomerase activity levels were significantly associated with shorter mean survival. In this regard, an inverse correlation between TL and telomerase activity in CLL was observed, where patients in early-stage disease had significantly longer telomeres than those at late stages [98, 99].

In plasma cell disorders, a few small studies on telomerase activity have been reported, in which telomerase level was not elevated in patients with MGUS, but it was increased in patients with MM and in all cases with plasma cell leukemia [74]. In our study we observed high *hTERT* expression in MM as well as in MGUS [83]. Analysis of results showed that both pathologies had a great heterogeneity in *hTERT* levels, with cases that over-expressed this gene and others with low mRNA expression. In agreement with our data, heterogeneity in telomerase activity has been previously referred for MM patients [74, 76] meanwhile the same behavior in MGUS was found for the first time by our group [83]. As for TL, an inverse correlation with *hTERT* levels was observed in both MGUS and MM, with shortest telomeres for patients that had the highest *hTERT* expression. Our results showing a group of MGUS patients with very short telomeres and high *hTERT* expression suggest a role for these parameters in the progression from MGUS to MM.

There have been a limited number of investigations evaluating the expression profile of telomere-associated proteins in lymphoid malignancies (Table 7).

Table 7. Telomere-associated genes reported in the literature

Subtype	Telomere-associated gene	References
CLL	*hTERT, DKC, TRF1,TRF2,hRAP1,TIN2,POT1,TPP1, hEST1A,MRE11,RAD50,Ku80,RPA1*	[77]
FL	*hTERT, TANK1, TRF1, TRF2, hPif1*	[78]
DLBCL	*hTERT, TANK1, TRF1, TRF2, hPif1*	[78]
BL	*hTERT, TANK1, TRF1, TRF2, hPif1*	[78]
MCL	*hTERT, TANK1, TRF1, TRF2, hPif1*	[78]
ATL	*TRF1, TRF2, TIN2*	[100]
MM	*hTERT, TANK1* *hTERT, TANK1, TRF1, TRF2*	[76] [83]
MGUS	*hTERT, TANK1, TRF1, TRF2*	[83]

CLL: chronic lymphocytic leukemia; FL: Follicular lymphoma; DLBCL: diffuse large B cell lymphoma; BL: Burkitt lymphoma; MCL: mantle cell lymphoma; MM: multiple myeloma; MGUS: monoclonal gammopathy of undetermined significance; ATL: adult T cell leukemia.

A number of studies have demonstrated increased mRNA levels of *TANK1*, *TRF2* and *hPif1* in high grade NHL [78], and of *TRF1*, *TRF2* and *TIN2* in adult T cell leukemia (ATL) [100]. In this entity, mRNA overexpression was associated with increased chromosomal instability and the risk of progressing to the acute stage of ATL. Another report has detected a low expression of many components involved in telomere elongations and capping in CLL [77].

In our study, a decrease of *TRF1* and an increase of *TRF2* and *TANK1* levels in patients respect to controls and in MM compared to MGUS were observed. The up-regulation of *TANK1* and its association with telomerase expression was previously observed by Xu et al. [74] in a small number of MM and plasma cell leukemia patients. Thus, our data confirm a previous observation in MM and, interestingly, showed a similar behavior for MGUS patients. In addition, we found that *TRF2* and *TANK1* levels increased with *hTERT* expression, showing a direct correlation between both *TRF2* and *TANK1* expression. These findings indicate an interaction between positive and negative factors in the maintenance of TL in plasma cell disorders and support a role for *TRF2* and *TANK1* up-regulation in disease progression.

Conclusion

We have analyzed TL in different B-cell disorders and its association with genetic characteristics and telomere-binding proteins expression. Our data confirm the heterogeneity of TRF length and *hTERT* expression in theses pathologies and showed changes in the expression of telomere maintenance genes, supporting the idea that telomere shortening leads to increase genomic instability and contributes to tumorigenesis.

These findings result of interest considering the new therapeutic approaches with diverse telomere-targeting inhibitors and suggest that a number of patients with B-cell malignancies may be particularly sensitive to telomere-damage drugs.

Acknowledgments

This work was supported by grants from the National Research Council (CONICET) and the National Agency of Scientific and Technical Promotion (ANPCyT).

References

[1] Blasco MA. (2005) Telomeres and human disease: aging, cancer and beyond. *Nat. Rev. Genet.*, 6, 611–22.

[2] Shay JW, Wright WE (2000) Hayflick, his limit, and cellular ageing. *Nat. Rev. Mol. Cell Biol.*, 1, 72-76.

[3] Denchi EL (2009) Give me a break: how telomeres suppress the DNA damage response. *DNA Repair,* 8:1118-26.

[4] Reddel RR (2000) The role of senescence and immortalization in carcinogenesis. *Carcinogenesis*, 21: 477-84.

[5] Shay JW, Werbin H, Wright WE (1996) Telomeres and telomerase in human leukemias. *Leukemia,* 10:1255-61.

[6] Hastie ND, Dempster M, Dunlop MG, Thompson AM, Green DK, Allshire RC (1990) Telomere reduction in human colorectal carcinoma and with ageing. *Nature*, 364: 866-8.

[7] Satoh H, Hiyama K, Takeda M, Awaya Y, Watanabe K, Ihara Y, Maeda H, Ishioka S, Yamakido M (1996) Telomere shortening in peripheral blood cells was related with aging but not with white blood cell count. *Jpn. J. Hum. Genet.*, 41: 413-7.

[8] Iwama H, Ohyashiki K, Ohyashiki JH, Hayashi S, Yahata N, Ando K, Toyama K, Hoshika A, Takasaki M, Mori M, Shay JW, (1998) Telomeric length and telomerase activity vary with age in peripheral blood cells obtained from normal individuals. *Hum. Genet.*, 102: 397-402.

[9] Lindsey J, McGill NI, Lindsey LA, Green DK, Cooke HJ (1991) In vivo loss of telomeric repeats with age in humans. *Mutat. Res.*, 256: 45-8.

[10] Vaziri H, Dragowska W, Allsopp RC, Thomas TE, Harley CB, Lansdorp PM (1994) Evidence for a mitotic clock in human hematopoietic stem cells: Loss of telomeric DNA with age. *Proc. Natl. Acad. Sci. USA.*, 91: 9857-60.

[11] Murnane JP, Sabatier L, Marder BA, Morgan WF (1994) Telomere dynamics in an immortal human cell line. *EMBO J.,* 13, 4953–62.

[12] Bryan TM, Englezou A, Gupta J, Bacchetti S, Reddel RR (1995) Telomere elongation in immortal human cells without detectable telomerase activity. *EMBO J.,* 14: 4240-8.

[13] Dunham MA, Neumann AA, Fasching CL, Reddel RR (2000) Telomere maintenance by recombination in human cells. *Nat. Genet.*, 26: 447-50.

[14] Counter CM, Gupta J, Harley CB, Leber B, Bacchetti S (1995) Telomerase activity in normal leukocytes and in hematologic malignancies. *Blood*, 85:2315-20.

[15] Shay JW, Zou Y, Hiyama E, Wright WE (2001) Telomerase and cancer. *Hum. Mol. Genet.*, 10: 677-85.

[16] Bryan TM, Englezou A, Dalla-Pozza L, Dunham MA, Reddel RR. (1997) Evidence for an alternative mechanism for maintaining telomere length in human tumors and tumor-derived cell lines. *Nat. Med.*, 3:1271-4.

[17] Shay JW, Bacchetti S (1997) A survey of telomerase activity in human cancer. *Eur. J. Cancer.*, 33: 787-91.

[18] Nabetani A, Ishikawa F (2011) Alternative lengthening of telomeres pathway: recombination-mediated telomere maintenance mechanism in human cells. *J. Biochem.*, 149: 5-14.

[19] Liu D, O'Connor MS, Qin J, Songyang Z. (2004) Telosome, a mammalian telomere-associated complex formed by multiple telomeric proteins *J. Biol. Chem.*, 279: 51338-42.

[20] de Lange T (2005) Shelterin: the protein complex that shapes and safeguards human telomeres. *Genes Dev.*, 19: 2100-10.

[21] Donate LE, Blasco MA (2010) Telomeres in cancer and ageing. *Philos. Trans. R. Soc. Lond B. Biol. Sci.*, 366: 76-84.

[22] van Steensel B, de Lange T (1997) Control of telomere length by the human telomeric protein TRF1. *Nature*, 385: 740-3.

[23] Kim SH, Beausejour C, Davalos AR, Kaminker P, Heo SJ, Campisi J. (2004) TIN2 mediates functions of TRF2 at human telomeres. *J. Biol. Chem.*, 279: 43799-804.

[24] Ye JZ, Hockemeyer D, Krutchinsky AN, Loayza D, Hooper SM, Chait BT, de Lange T (2004) POT1-interacting protein PIP1: A telomere length regulator that recruits POT1 to the TIN2/TRF1 complex. *Genes. Dev.*, 18: 1649-54.

[25] Chen LY, Liu D, Songyang Z (2007) Telomere maintenance through spatial control of telomeric proteins. *Mol. Cell Biol.*, 27: 5898-909.

[26] Xin H, Liu D, Wan M, Safari A, Kim H, Sun W, O'Connor MS, Songyang Z (2007) TPP1 is a homologue of ciliate TEBP-beta and interacts with POT1 to recruit telomerase. *Nature*, 445: 559-62.

[27] Li B, Oestreich S, de Lange T (2000) Identification of human Rap1: implications for telomere evolution. *Cell*, 101: 471-83.

[28] Li B, de Lange T (2003) Rap1 affects the length and heterogeneity of human telomeres. *Mol. Biol. Cell*, 14: 5060-8.

[29] Celli GB, de Lange T (2005) DNA processing is not required for ATM-mediated telomere damage response after TRF2 deletion. *Nat. Cell Biol.*, 7: 712-8.

[30] Verdun RE, Karlseder J (2007) Replication and protection of telomeres. *Nature*, 447: 924-31.

[31] Smith S, Giriat I, Schmitt A, de Lange T (1998) Tankyrase, a poly (ADP-ribose) polymerase at human telomeres. *Science*, 282: 1484-7.

[32] Smith S, de Lange T (2000) Tankyrase promotes telomere elongation in human cells. *Curr. Biol.*, 10: 1299-1302.

[33] Karlseder J, Smogorzewska A, De Lange T (2002) Senescence induced by altered telomere state, not telomere loss. *Science*, 295: 2446-9.

[34] Denchi EL, de Lange T (2007) Protection of telomeres through independent control of ATM and ATR by TRF2 and POT1. *Nature*, 448: 1068-71.

[35] DePinho RA, Polyak K (2004) Cancer chromosomes in crisis. *Nat. Genet.*, 36: 932-4.

[36] Deng Y, Chan SS, Chang S (2008) Telomere dysfunction and tumor suppression: the senescence connection. *Nature*, 8: 450-8.

[37] Maser RS, DePinho RA (2002) Keeping telomerase in its place. *Nat. Med.*, 8: 934-6.

[38] Hahn WC (2005) Telomere and telomerase dynamics in human cells. *Curr. Mol. Med.*, 5: 227-31.

[39] Blasco MA, Lee HW, Hande MP, Samper E, Lansdorp PM, DePinho RA, Greider CW (1997) Telomere shortening and tumor formation by mouse cells lacking telomerase *RNA. Cell*, 91: 25-34.

[40] Artandi SE, Chang S, Lee SL, Alson S, Gottlieb GJ, Chin L, DePinho RA (2000) Telomere dysfunction promotes non-reciprocal translocations and epithelial cancers in mice. *Nature*, 406: 641-5.

[41] Swerdlow SH, Campo E, Harris NL, Jaffe ES, Pileri SA, Stein H, Thiele J, Vardiman JW (eds). WHO classification of tumors of Haematopoietic and Lymphoid Tissues. Lyon: *IARC Press*,; 2008.

[42] Chiu BC-H. Epidemiology of non-Hodgkin´s lymphomas. In: Vose JM, Chiu BC-H, Cheson BD, Dancey J, Wright J. Update on epidemiology and therapeutics for non-Hodgkin´s lymphoma. *Am. Soc. Hematol.*, Education Program Book. 2002, pp. 241-5.

[43] Cleary ML, Cklar J (1985) Nucleotide sequence of a t(14;18) chromosome breakpoint in follicular lymphoma and demonstration of a breakpoint cluster region near a

transcriptionally active locus on chromosome 18. *Proc. Natl. Acad. Sci. USA.*, 82: 7439-41.

[44] Clearly ML, Galili N, Sklar J (1986) Detection of a second t(14;18) breakpoint cluster region in human follicular lymphomas. *J. Exp. Med.*, 164: 315-20.

[45] Saintigny Y, Dumay A, Lambert S, Lopez BS (2001) A novel role for the Bcl-2 protein family: specific suppression of the RAD51 recombination pathway. *EMBO J.*, 10: 2596-607.

[46] Youn CK, Cho HJ, Kim SH, Kim HB, Kim MH, Chang IY, Lee JS, Chung MH, Hahm KS, You HJ (2005) Bcl-2 expression suppresses mismatch repair activity through inhibition of E2F transcriptional activity. *Nat. Cell Biol.*, 7:137-47.

[47] Dave BJ, Weisenburger DD, Higgins CM, Pickering DL, Hess MM, Chan WC, Sanger WG (2004) Cytogenetics and fluorescence in situ hybridization studies of diffuse large B-cell lymphoma in children and young adults. *Cancer Genet. Cytogenet.*, 153: 115-21.

[48] Cerretini R, Noriega MF, Narbaitz M, Slavutsky I (2006) New chromosome abnormalities and lack of BCL-6 gene rearrangements in Argentinean diffuse large B-cell lymphoma. *Eur. J. Haematol.*, 76: 284-93.

[49] Alizadeh AA, Eisen MB, Davis RE, Ma C, Lossos IS, Rosenwald A, Boldrick JC, Sabet H, Tran T, Yu X, Powell JI, Yang L, Marti GE, Moore T, Hudson J Jr, Lu L, Lewis DB, Tibshirani R, Sherlock G, Chan WC, Greiner TC, Weisenburger DD, Armitage JO, Warnke R, Levy R, Wilson W, Grever MR, Byrd JC, Botstein D, Brown PO, Staudt LM (2000) Distinct types of diffuse large B-cell lymphoma identified by gene expression profiling. *Nature*, 403: 503-11.

[50] Rosenwald A, Wright G, Chan WC, Connors JM, Campo E, Fisher RI, Gascoyne RD, Müller-Hermelink HK, Smeland EB, Giltnane JM, Hurt EM, Zhao H, Averett L, Yang L, Wilson WH, Jaffe ES, Simon R, Klausner RD, Powell J, Duffey PL, Longo DL, Greiner TC, Weisenburger DD, Sanger WG, Dave BJ, Lynch JC, Vose J, Armitage JO, Montserrat E, López-Guillermo A, Grogan TM, Miller TP, LeBlanc M, Ott G, Kvaloy S, Delabie J, Holte H, Krajci P, Stokke T, Staudt LM. Lymphoma/Leukemia Molecular Profiling Project (2002) The use of molecular profiling to predict survival after chemotherapy for diffuse large-B-cell lymphoma. *N. Engl. J. Med.*, 346: 1937-47.

[51] Rosenwald A, Wright G, Leroy K, Yu X, Gaulard P, Gascoyne RD, Chan WC, Zhao T, Haioun C, Greiner TC, Weisenburger DD, Lynch JC, Vose J, Armitage JO, Smeland EB, Kvaloy S, Holte H, Delabie J, Campo E, Montserrat E, Lopez-Guillermo A, Ott G, Müller-Hermelink HK, Connors JM, Braziel R, Grogan TM, Fisher RI, Miller TP, LeBlanc M, Chiorazzi M, Zhao H, Yang L, Powell J, Wilson WH, Jaffe ES, Simon R, Klausner RD, Staudt LM (2003) Molecular diagnosis of primary mediastinal B cell lymphoma identifies a clinically favorable subgroup of diffuse large B cell lymphoma related to Hodgkin lymphoma. *J. Exp. Med.*, 198: 851-62.

[52] Camacho FI, Algara P, Rodríguez A, Ruíz-Ballesteros E, Mollejo M, Martínez N, Martínez-Climent JA, González M, Mateo M, Caleo A, Sánchez-Beato M, Menárguez J, García-Conde J, Solé F, Campo E, Piris MA (2003) Molecular heterogeneity in MCL defined by the use of specific VH genes and the frequency of somatic mutations. *Blood*, 101: 4042-6.

[53] Kienle D, Kröber A, Katzenberger T, Ott G, Leupolt E, Barth TF, Möller P, Benner A, Habermann A, Müller-Hermelink HK, Bentz M, Lichter P, Döhner H, Stilgenbauer S (2003) VH mutation status and VDJ rearrangement structure in mantle cell lymphoma:

correlation with genomic aberrations, clinical characteristics, and outcome. *Blood*, 102: 3003-9.

[54] Hallek M, Bergsagel PL, Anderson KC (1989) Multiple myeloma: increasing evidence for a multistep transformation process. *Blood*, 91: 3-21.

[55] Kyle RA, Rajkumar SV (2004) Multiple myeloma. *N. Engl. J. Med.*, 351: 1860-73.

[56] Zhan F, Sawyer J, Tricot G. (2006) The role of cytogenetics in myeloma. *Leukemia*, 20: 1484-6.

[57] Kyle RA, Therneau TM, Rajkumar SV, Offord JR, Larson DR, Plevak MF, Melton LJ 3rd (2002) A long term study of prognosis in monoclonal gammopathy of undetermined significance. *N. Engl. J. Med.*, 346: 564-9.

[58] Chiorazzi N, Rai KR, Ferrarini M. (2005) Chronic lymphocytic leukemia. *N. Eng. J. Med.*, 352: 804-15.

[59] Döhner H, Stilgenbauer S, Benner A, Leupolt E, Kröber A, Bullinger L, Döhner K, Bentz M, Lichter P (2000) Genomic aberrations and survival in chronic lymphocytic leukemia. *N. Engl. J. Med.*, 343:1910-6.

[60] Dewald GW, Brockman SR, Paternoster SF, Bone ND, O'Fallon JR, Allmer C, James CD, Jelinek DF, Tschumper RC, Hanson CA, Pruthi RK, Witzig TE, Call TG, Kay NE (2003) Chromosome anomalies detected by interphase fluorescence in situ, hybridization: correlation with significant biological features of B-cell chronic lymphocytic leukaemia. *Br. J. Haematol.*, 121: 287-95.

[61] Chena C, Avalos JS, Bezares RF, Arrossagaray G, Turdó K, Bistmans A, Slavutsky I (2008) Biallelic deletion 13q14.3 in patients with chronic lymphocytic leukemia: cytogenetic, FISH and clinical studies. *Eur. J. Haematol.*, 81: 94-9.

[62] Mayr C, Speicher MR, Kofler DM, Buhmann R, Strehl J, Busch R, Hallek M, Wendtner CM (2006) Chromosomal translocations are associated with poor prognosis in chronic lymphocytic leukemia. *Blood*, 107: 742-51.

[63] Van Den Neste E, Robin V, Francart J, Hagemeijer A, Stul M, Vandenberghe P, Delannoy A, Sonet A, Deneys V, Costantini S, Ferrant A, Robert A, Michaux L (2007) Chromosomal translocations independently predict treatment failure, treatment-free survival and overall survival in B-cell chronic lymphocytic leukemia patients treated with cladribine. *Leukemia*, 21: 1715-22.

[64] Ricca I, Rocci A, Drandi D, Francese R, Compagno M, Lobetti Bodoni C, De Marco F, Astolfi M, Monitillo L, Vallet S, Calvi R, Ficara F, Omedè P, Rosato R, Gallamini A, Marinone C, Bergui L, Boccadoro M, Tarella C, Ladetto M (2007) Telomere length identifies two different prognostic subgroups among VH-unmutated B-cell chronic lymphocytic leukemia patients. *Leukemia*, 21:697-705.

[65] Rossi D, Lobetti Bodoni C, Genuardi E, Monitillo L, Drandi D, Cerri M, Deambrogi C, Ricca I, Rocci A, Ferrero S, Bernocco E, Capello D, De Paoli L, Bergui L, Boi M, Omedè P, Massaia M, Tarella C, Passera R, Boccadoro M, Gaidano G, Ladetto M (2009) Telomere length is an independent predictor of survival, treatment requirement and Richter's syndrome transformation in chronic lymphocytic leukemia. *Leukemia*, 23: 1062-72.

[66] Roos G, Kröber A, Grabowski P, Kienle D, Bühler A, Döhner H, Rosenquist R, Stilgenbauer S (2008) Short telomeres are associated with genetic complexity, high-risk genomic aberrations, and short survival in chronic lymphocytic leukemia. *Blood*, 111: 2246-52.

[67] Ladetto M, Compagno M, Ricca I, Pagano M, Rocci A, Astolfi M, Drandi D, di Celle PF, Dell'Aquila M, Mantoan B, Vallet S, Pagliano G, De Marco F, Francese R, Santo L, Cuttica A, Marinone C, Boccadoro M, Tarella C (2004) Telomere length correlates with histopathogenesis according to the germinal center in mature B-cell lymphoproliferative disorders. *Blood*, 103: 4644-9.

[68] Walsh SH, Grabowski P, Berglund M, Thunberg U, Thorsélius M, Tobin G, Aleskog A, Karlsson K, Sundström C, Laurell A, Enblad G, Rosenquist R, Roos G (2007) Telomere length and correlation with histopathogenesis in B-cell leukemias/lymphomas. *Eur. J. Haematol.*, 78:283-9.

[69] Cottliar AS, Panero J, Pedrazzini E, Noriega MF, Narbaitz M, Rodríguez A, Slavutsky I. (2009) Analysis of telomere length in mantle cell lymphoma. *Eur. J.Haematol.*, 83: 433-8.

[70] Norrback KF, Dahlenborg K, Carlsson R, Roos G (1996) Telomerase activation in normal B lymphocytes and non-Hodgkin's lymphomas. *Blood*, 88: 222-9.

[71] Ely SA, Chadburn A, Dayton CM, Cesarman E, Knowles DM (2000) Telomerase activity in B-cell non- Hodgkin lymphoma. *Cancer* 89: 445-52.

[72] Remes K, Norrback KF, Rosenquist R, Mehle C, Lindh J, Roos G (2000) Telomere length and telomerase activity in malignant lymphomas at diagnosis and relapse. *Br. J. Cancer*, 82: 601-7.

[73] Lin Z, Lim S, Viani MA, Sapp M, Lim MS (2001) Down-regulation of telomerase activity in malignant lymphomas by radiation and chemotherapeutic agents. *Am. J. Pathol.*, 159: 711-9.

[74] Xu D, Zheng C, Bergenbrant S, Holm G, Björkholm M, Yi Q, Gruber A (2001) Telomerase activity in plasma cell dyscrasias. *Br. J. Cancer*, 84: 621-5.

[75] Cottliar A, Pedrazzini E, Corrado C, Engelberger MI, Narbaitz M, Slavutsky I (2003) Telomere shortening in patients with plasma cell disorders. *Eur. J. Haematol.*, 71: 334-40.

[76] Wu KD, Orme LM, Shaughnessy J Jr, Jacobson J, Barlogie B, Moore MA (2003) Telomerase and telomere length in multiple myeloma: correlations with disease heterogeneity, cytogenetic status, and overall survival. *Blood*, 101: 4982-9.

[77] Poncet D, Belleville A, t'kint de Roodenbeke C, Roborel de Climens A, Ben Simon E, Merle-Beral H, Callet-Bauchu E, Salles G, Sabatier L, Delic J, Gilson E (2008) Changes in the expression of telomere maintenance genes suggest global telomere dysfunction in B-chronic lymphocytic leukemia. *Blood*, 111: 2388-91.

[78] Klapper W, Krams M, Qian W, Janssen D, Parwaresch R (2003) Telomerase activity in B-cell non-Hodgkin lymphomas is regulated by hTERT transcription and correlated with telomere-binding protein expression but uncoupled from proliferation. *Br. J. Cancer*, 89: 713-9.

[79] Hu LH, Chen FH, Li YR, Wang L (2004) Real-time determination of human telomerase reverse transcriptase mRNA in gastric cancer. *World J. Gastroenterol.*, 10: 3514-7.

[80] ISCN. An International System for Human Cytogenetic Nomenclature. Shaffer LG, Slovak ML, Campbell LJ (eds), Karger, Basel (2009).

[81] Noriega MF, De Brasi C, Narbaitz M, Slavutsky I (2004) Incidence of BCL-2 gene rearrangements in Argentinean non-Hodgkin lymphoma patients: increased frequency of breakpoints outside of MBR and MCR. *Blood Cells Mol. Dis.*, 32: 232-9.

[82] Cottliar HAS, Noriega MF, Narbaitz M, Rodríguez A, Slavutsky IR (2006) Association between telomere length and BCL2 gene rearrangements in low- and high-grade non-Hodgkin lymphomas. *Cancer Genet. Cytogenet.*, 171: 1-8.

[83] Panero J, Arbelbide J, Fantl DB, Rivello HG, Kohan D, Slavutsky I (2010) Altered mRNA expression of telomere-associated genes in monoclonal gammopathy of undetermined significance and multiple myeloma. *Mol. Med.*, 16: 471-8.

[84] Wu X, Amos CI, Zhu Y, Zhao H, Grossman BH, Shay JW, Luo S, Ki Hong W, Spitz MR. (2003) Telomere dysfunction: a potential cancer predisposition factor. *J. Nat. Cancer Inst.*, 95:1211-8.

[85] Fonseca R, Bergsagel PL, Drach J, Shaughnessy J, Gutierrez N, Stewart AK, Morgan G, Van Ness B, Chesi M, Minvielle S, Neri A, Barlogie B, Kuehl WM, Liebisch P, Davies F, Chen-Kiang S, Durie BG, Carrasco R, Sezer O, Reiman T, Pilarski L, Avet-Loiseau H; International Myeloma Working Group (2009) International Myeloma Working Group molecular classification of multiple myeloma: spotlight review. *Leukemia*, 23: 2210-21.

[86] Stella F, Pedrazzini E, Rodríguez A, Baialardo E, Kusminsky G, Arbelbide J, Fantl DB, Slavutsky I (2011) New recurrent chromosome alterations in patients with multiple myeloma and plasma cell leukemia. *Cytogenet. Genome. Res.*, 134: 249-59.

[87] Lin TT, Letsolo BT, Jones RE, Rowson J, Pratt G, Hewamana S, Fegan C, Pepper C, Baird DM (2010) Telomere dysfunction and fusion during the progression of chronic lymphocytic leukemia: evidence for a telomere crisis. *Blood* 116: 1899-907.

[88] Horsman DE, Okamoto I, Ludkovsky O, Le N, Harder L, Gesk S, Siebert R, Chhanabahai M, Sehn L, Connors JM, Gascoyne RD (2003) Follicular lymphoma lacking the t(14;18)(q32;q21): identification of two disease subtypes. *Br. J. Haematol.*, 120: 424-33.

[89] Nanjangud G, Naresh K, Teruya-Feldstein J, Olshen AB, Wilton AS, Parsa NZ, Rao PH, Filippa DA, Hegde A, Hedvat CV, Jhanwar SC, Mehra S, Johansson B, Mitelman F, Zelenetz AD, Chaganti RSK (2003) Cytogenetic analysis of follicular lymphoma: Chromosome instability associated with t(14;18)(q32;q21) positive and negative subsets, histology and transformations. *Blood*, 102:847a.

[90] Nanjangud G, Rao PH, Teruya-Feldstein J, Donnelly G, Qin J, Mehra S, Jhanwar SC, Zelenetz AD, Chaganti RS (2007) Molecular cytogenetic analysis of follicular lymphoma (FL) provides detailed characterization of chromosomal instability associated with the t(14;18)(q32;q21) positive and negative subsets and histologic progression. *Cytogenet. Genome Res.*, 118: 337-44.

[91] Jardin F, Gaulard P, Buchonnet G, Contentin N, Leprêtre S, Lenain P, Stamatoullas A, Picquenot JM, Duval C, Parmentier F, Tilly H, Bastard C (2002) Follicular lymphoma without t(14;18) and with BCL-6 rearrangement: a lymphoma subtype with distinct pathological, molecular and clinical characteristics. *Leukemia*, 16: 2309-17.

[92] Montoto S, López-Guillermo A, Colomer D, Esteve J, Bosch F, Ferrer A, Villamor N, Moreno C, Campo E, Montserrat E (2003) Incidence and clinical significance of bcl-2/IgH rearrangements in follicular lymphoma. *Leuk. Lymphoma*, 44: 71-6.

[93] Leich E, Salaverria I, Bea S, Zettl A, Wright G, Moreno V, Gascoyne RD, Chan WC, Braziel RM, Rimsza LM, Weisenburger DD, Delabie J, Jaffe ES, Lister A, Fitzgibbon J, Staudt LM, Hartmann EM, Mueller-Hermelink HK, Campo E, Ott G, Rosenwald A

(2009) Follicular lymphomas with and without translocation t(14;18) differ in gene expression profiles and genetic alterations. Blood, 114: 826-34.

[94] De Cian A, Lacroix L, Douarre C, Temime-Smaali N, Trentesaux C, Riou JF, Mergny JL (2008) Targeting telomeres and telomerase. *Biochimie*, 90:131-5.

[95] Folini M, Venturini L, Cimino-Reale G, Zaffaroni N (2011) Telomeres as targets for anticancer therapies. *Expert Opin. Ther. Targets*, 15: 579-93.

[96] Bechter OE, Eisterer W, Pall G, Hilbe W, Kuhr T, Thaler J (1998) Telomere length and telomerase activity predict survival in patients with B cell chronic lymphocytic leukemia. *Cancer Res.*, 58: 4918- 22.

[97] Trentin L, Ballon G, Ometto L, Perin A, Basso U, Chieco-Bianchi L, Semenzato G, De Rossi A (1999) Telomerase activity in chronic lymphoproliferative disorders of B-cell lineage. *Br. J. Haematol.*, 106: 662-668.

[98] Damle RN, Batliwalla FM, Ghiotto F, Valetto A, Albesiano E, Sison C, Allen SL, Kolitz J, Vinciguerra VP, Kudalkar P, Wasil T, Rai KR, Ferrarini M, Gregersen PK, Chiorazzi N (2004) Telomere length and telomerase activity delineate distinctive replicative features of the B-CLL subgroups defined by immunoglobulin V gene mutations. *Blood*, 103: 375-82.

[99] Terrin L, Trentin L, Degan M, Corradini I, Bertorelle R, Carli P, Maschio N, Bo MD, Noventa F, Gattei V, Semenzato G, De Rossi A (2007) Telomerase expression in B-cell chronic lymphocytic leukemia predicts survival and delineates subgroups of patients with the same igVH mutation status and different outcome. *Leukemia*, 21: 965-72.

[100] Bellon M, Datta A, Brown M, Pouliquen JF, Couppie P, Kazanji M, Nicot C (2006) Increased expression of telomere length regulating factors TRF1, TRF2 and TIN2 in patients with adult T-cell leukemia. *Int. J. Cancer*, 119: 2090-7.

In: Advances in Genetics Research. Volume 12
Editor: Kevin V. Urbano

ISBN: 978-1-63321-351-7

Chapter 6

TELOMERASE AND TELOMERE MAINTENANCE IN THE BLIGHT FUNGUS *USTILAGO MAYDIS*

Patricia Sánchez-Alonso[2], Candelario Vázquez-Cruz[2], Guillermo Horta-Valerdi[2] and Plinio Guzmán[1]

[1]Departamento de Ingeniería Genética de Plantas, Centro de Investigación y de Estudios Avanzados, Unidad Irapuatokm 9.6 Libramiento Norte, Irapuato, Gto,. Mexico

[2]Centro de Investigaciones Microbiológicas, Instituto de Ciencias, Benemérita Universidad Autónoma de Puebla, Puebla, Mexico

ABSTRACT

Telomeres, the ribonucleoprotein complexes located at the chromosomal termini, are committed to cap and protect the endmost chromosomal sequences from being recognized as double strand breaks by the DNA damage repair machinery. Telomeres are synthesized by the functions of the telomerase, a ribonucleoproteic enzyme-complex with reverse transcriptase properties that maintains telomere length homeostasis in eukaryotic cells. In *Ustilago maydis*, a basidiomycetous phytopathogen that infects many plant species of the *Zea* genus, the telomere is composed af a TTAGGG repeated motif, which is commonly found in vertebrates and other fungal species. Both the telomere organization and the characterization of the telomerase catalytic subunit have been characterized in this fungus. Adjacent to the telomere are mosaics of sequences, which are largely species-specific and commonly termed telomere-associated sequences (TAS), The TAS are not fully understood and require further characteriation. *UTASa*, a class of TAS in *U. maydis,* harbors an open reading frame with homology to RecQ enzymes. Those sequences resemble a long non-coding RNA (lncRNA) and exhibit a variable expression pattern during the cell cycle in synchronous cell cultures. We propose that this type of RecQ-like sequence may be part of a surveillance mechanism that operates in the absence of telomerase.

Introduction

Telomere, the nucleoprotein complex located at the endmost chromosomal terminus, is one of the essential *cis*-acting structural elements of the chromosome. Telomeres are composed of DNA, proteins, and RNA (Azzalin & Ligner, 2008). They sense and act on signaling of cell cycle progression, and prevent the recognition of chromosome ends as double strand breaks by DNA damage repair systems. This mechanism prevents the uncontrolled resection of terminal DNA and enhances the genomic stability and viability of eukaryotic cells (Chan & Blackburn, 2004).

In most eukaryotic species, telomeric DNA is composed of short GC rich DNA-repeat motifs that are tandemly arranged and oriented 5' to 3' toward the chromosome end. Telomeres are sequences of dynamic length, even on chromosomes from the same cell, but have an average size that is characteristic for each species. The telomere has two domains. The first one comprises tracts of double stranded telomere repeats that have an average length related to the donor species (i.e., 150 Kbp for mouse and 20 bp for *Oxytricha*), donor age, or tissue type in higher eukaryotes (Harley et al., 1990; Kipling & Cooke, 1990; Klobutcher et al., 1981). The second domain is a protruding overhang from the 3' G-strand that is a few bases in length, and has been found at the furthermost telomere end in nearly all organisms analyzed. This single-stranded domain of the telomere has important roles in telomere maintenance as described later (Raices et al., 2008, and references therein). The telomere is bound to a variety of telomere-specific proteins that belong to either the family of factors involved in *D*NA *D*amage *R*esponse (DDR), or the protective protein complex termed shelterin or telosome in higher eukaryotes and yeast, respectively (Liu et al., 2004; Palm & de Lange, 2008; Wright et al., 1992). In *S. cerevisiae*, two groups of three proteins protect the chromosome ends. In one group, Rap1, the most abundant telomere-binding protein, binds to the double-stranded domain of the telomere repeat and recruits *R*ap1 *I*nteracting *F*actors 1 and 2 (Rif1 and Rif2) or nucleates the Sir2/3/4 complex to initiate the silenced state of chromatin. However, the binding of repressor/activator protein 1 (Rap1p) to Rif1/Rif2 or the Sir2/3/4 complex is not sufficient for protecting telomere function (Bourns et al., 1998; Hardy et al., 1992). In the second group, a heterotrimeric protein complex composed of Cdc13, Snt1, and Ten1 (collectively called the CST complex) binds to the ssDNA overhang and exerts a protective action that is related to shelterin in mammals (Price et al., 2010). In vertebrates, the shelterin complex is composed by six proteins; TRF1, TRF2, TIN2, Rap1, TPP1, and POT1 (Liu et al., 2004). Of these proteins, TRF1, TRF2, and POT1 bind telomere repeats directly, either on double- (TRF1 and TRF2) or on single-stranded DNA (POT1), whereas TIN2, RAP1 and TPP1 are recruited by these telomere binding proteins (Palm & de Lange, 2008). Shelterin and telosome protective complexes are functional and structurally related. Each complex contains telomere-specific proteins and both are important for chromosome end protection and telomere length regulation (Price et al., 2010). In vertebrates, shelterin and the DDR proteins, or shelterin-associated factors, repress the DNA damage response that could potentially occur at otherwise free-ends of telomeres. Moreover, the debate about the presence of DDR proteins as telomere-associated factors and the mechanism by which they avoid the recognition of the chromosome terminus as *D*ouble *S*trand *B*reaks (DBS) continues even today (Palm & de Lange, 2008). The CST complex has a central role in the control of telomere replication and chromosome end protection in *S. cerevisiae*, and its components are not exclusively expressed in this fungus, they are conserved in plants and vertebrates and coexist with shelterin proteins. Accordingly, a new nucleoprotein structure engaged in telomere protection is currently being explored that takes into account the new proteins as well as the six proteins initially reported as shelterin (Giraud-Panis et al., 2010). The 3' G-strand overhang is also important for telomere synthesis and provides the substrate for telomere lengthening at the G-rich strand. This overhang is essential for t-loop stabilization, which is the structure that provides protection to chromosome ends through the prevention of recognition as broken DNA by DDR mechanisms.

T-loop structures are the most stable shape of telomeres and naturally occur in higher and lower eukaryotes from different taxa, including human, plant, protozoan, and under certain conditions, fungus (Cesare et al., 2003, 2008; Griffith et al., 1999; Munoz-Jordan et al., 2001). T-loop structures assemble by invasion of the single-stranded 3' G-rich overhang into the internal double-stranded regions of the telomere repeat. This process displaces the internal G-rich strand and causes the creation of a half Holliday junction, which is a hallmark structure in t-loop formation and t-circle generation (Griffith et al., 1999). On other hand, the 3' G-rich overhang requires the controlled resection of the complementary 5' strand for its formation. This process is poorly understood in mammals, where the overhang formation seems to be regulated by multiple factors and, at least in part, by cell cycle prosecution (Dai et al., 2010). In *S. cerevisiae*, the 3' G-rich overhang is produced during late S phase and its formation is cell cycle controlled by resection of the 5'-strand from the telomeric repeat, which is mediated by the heterotrimeric complex Mre11, Rad50, and Xrs2 (MRX) (Vodenicharov & Wellinger, 2006). MRX is a multifaceted protein composed by Mre11 exonuclease activity, the SMC-related protein Rad50, and the Nbs1-related gene *Xrs2* in yeast. The MRX complex is central in telomere end resection, protection against recombination, and DBS response and damage checkpoint activation. In addition, it has been recognized that MRN has an important role in recognizing and sensing DBS, cell-cycle checkpoint activation, and regulation of chromatin remodeling near DBS in humans (Williams *et al.*, 2007). T-loops and t-circles are important structures for telomere maintenance by telomerase or the alternative recombination-based (ALT) pathways of telomere maintenance in the absence of telomerase (Griffith et al., 1999; Wang et al., 2004). T-loops prevent the recognition of telomeres as double strand breaks by DDR mechanisms by regulating telomere DNA lengthening and providing access to telomerase. This protection occurs when t-loops change from a linear and extendable conformation to a loop-shaped, non-extendable conformation (Teixeira et al., 2004). Invasion of the ssDNA strand into internal regions of the telomere seems to be absent in *S. cerevisiae*, and instead structures that are folded back have been suggested to provide protective caps for chromosome ends. That hypothesis was based on the absence of Holliday junctions as well as genetic evidence for protein interactions between the telomere and telomere-associated sequences (TAS) (de Bruin et al., 2001). Finally, although 3' G-strand overhangs were previously considered to be the only type of ssDNA overhang at chromosome termini, naturally occurring 3' G-strand and 5' C-strand overhangs have been recently reported in *Caenorhabditis elegans* as well as in some malignant mammalian cells as the result of recombination-based mechanisms for telomere maintenance (Oganesian & Karlseder, 2011; Raices et al., 2007).

THE *U. MAYDIS* CHROMOSOMAL ENDS

Telomeres in the basidiomycetous fungus *U. maydis* were reported in 1996 by Guzman and Sanchez (1996), and the telomere dynamics are somewhat understood due to the recent launch of this fungus as a model system for telomere metabolism. The telomeres in *U. mayis* are composed of TTAGGG repeats, which is the same repeated motif found in most filamentous fungi and vertebrates. Moreover, this repeat is canonical for those synthesized by the hetero-multimeric ribonucleoprotein enzyme responsible for telomere maintenance, telomerase. The mean length of *U. maydis* telomeres is approximately 240 bp in many wild isolates and laboratory strains, but the length can range from 50 to 1000 bp. Two telomere clones, UT1 and UT3, have been isolated, and the lengths of these regions as determined with the *Sau*3AI restriction enzyme were 595 and 6007 bp, respectively. These clones had 37-39 telomere repeats, which left a remainder of approximately 370 bp of telomere-associated sequence that was named *rumT* (Figure 1A). *RumT* is a composite of short G+C and A+T rich sequences. In the UT1 clone, the fragment comprises two regions that are delimited by a well-conserved *Pst*I recognizing site found in most of the telomere clones. The first region, which preserves the telomere repeat, was identified as the 357 bp endmost fragment (EF), and the remaining 238 bp of sequence were termed the endmost associated fragment (EAF) from the *U. maydis* telomere. EF has been useful for determining the telomere length as terminal restriction fragment (TRF) and EAF was proposed as a fingerprint probe for the analysis of genetic diversity in this fungus. A strong relationship between such fingerprint pattern and geographic origin was detected for *U. maydis* isolates (see right panel on Figure 3, and Sanchez Alonso et al., 1996).

No experimental characterization has been conducted on telomere binding or telomere-associated proteins in this fungus; however, we previously predicted some components based on the sequence alignment of *U. maydis* genome sequences with well characterized proteins from publicly available sequence databases (http://www.broad.mit.edu). Based in this analysis, we predicted that TRF1 and POT1, which are two of the most conserved proteins in several model organisms, were shelterin components (Sanchez-Alonso & Guzman, 2008); however, it is unknown whether newly identified telomere binding proteins share homology to the remaining four proteins of the shelterin complex, or if they would be homologues of those of the yeast telosome, since *U. maydis* is a lower eukaryote. Telomere-associated proteins from the DDR pathway in this fungus were also predicted, including the MRN complex, Ku70/80 heterodimer, and the RecQ-related helicase, *UmBLM* (Mao et al. 2009). In addition, ATM and ATR kinases of the DDR were also predicted in *U. maydis* by sequence alignment with public databases of protein sequences.

Middle-repeated sequences, known as telomere-associated sequences (TAS), are collinear with telomeres and comprise mosaics of interspersed sequences occurring at chromosome ends of most of the studied organisms to date (Louis & Haber 1992). In general, TAS includes non-coding sequences, retrotransposons, retroviruses, genes, and remnants of these elements, all of which are generally species-specific and diverse in their nature (Biessman & Mason, 1994; Louis et al., 1994). In the overall organization, subtelomeric sequences of *U. maydis* resemble those of *S. cerevisiae,* whereby there are two main types of telomere-associated sequences located contiguous to telomere repeats named Y' and X elements. The Y' element is a conserved TAS of the subtelomeric domain that is present in two major classes: Y' long, which is 6.7 kb in length, and Y' short, which is 5.2 kb in length. Y' can be found tandemly arranged and occurs in 0-4 copies at each single telomere (Chan & Tye, 1983; Louis & Haber, 1992). A repeated element X is adjacent to Y' sequences and proximal to the chromosome end. X elements are larger and more heterogeneous than Y elements and consist of a combination of core X and four other partially conserved *s*ub*t*elomeric *r*epeated elements named STR-A, -B, -C, and -D. Core X is a variable sequence of up to 559 bp that is conserved in most telomeres of yeast. Both the Y' and the X elements frequently contain autonomously replicating sequences (ARS) (Chan & Tye, 1983; Louis et al., 1994). At the junction of X and Y' elements, tracts of $(TG_{1-3})_n$ telomere sequences can be found that enable Y' elements to expand through recombination-based mechanisms (Louis et al., 1992). Two main telomere-associated sequences have been reported in *U. maydis*, including *UTASa,* which is a highly conserved element that exhibits greater than 96% homology with adjacent regions of the telomere. This subtelomeric element, which is currently being investigated, harbors an open reading frame of variable length on its 5' side and contains sequences that are homologous to RecQ-like helicases. The second TAS, *UTASb,* is a less conserved element based on the overall nucleotide sequence, but exhibits a conserved 830 bp. This element can be adjacent to telomere repeats or collinear with *UTASa*, and seems to possess ORF remnants that encode truncated proteins with unknown functions (depicted in Figure 1A). At the border between the telomere and subtelomeric sequences, the *rumT* element is always present (Sanchez-Alonso & Guzman, 1998). This well conserved element can also be found scattered throughout subtelomeric regions in most isolates and laboratory strains tested to date. An alignment of *rumT* sequences identified a variable region in all of the clones at similar positions (Figure 1B). *RumT* is amplified in survivors of telomerase-deficient strains of *U. maydis*, which suggests an important role in telomere function in the absence of telomerase (Sanchez-Alonso, unpublished). Although previous studies have provided a wealth of information, it is still currently difficult to assign a particular function to subtelomeric sequences. In different model systems, TAS have been thought to provide insulation for transcriptional silencing (Fourel et al., 1999), be preferential sites for recombinational rearrangements or sequence homogenization, or provide telomere functions in the absence of telomerase (Lundblad & Blackburn, 1993; Horowitz et al., 1984). Moreover, TAS are currently hypothesized to be active genetic elements with the capability to be transcribed and involved in the assembly of silent chromatin.

TAS have great plasticity, are tolerant of hypermutability, and are subject to chromosomal rearrangements at high frequency. These features can contribute to the generation of new alleles and rearrangements of sequence elements and genes that are harbored in subtelomeric regions (van Overveld et al., 2000). These characteristics account for the rapid evolution of contingency genes, which is a term coined by Moxon et al. (1998) that describes genes whose expression confer a phenotype that repairs growth defects due to rapid changes in the environment. These genes acquire variability due to the high mutagenic rate in the genetic context where they are naturally harbored. Contingency genes are important in antigenic variation in trypanosomatid parasites, carbohydrate metabolism in yeast, in lignin degradation in *Pleurotus ostreatus*, and other functions (Barry et al 2003; Perez et al 2009). In *U. maydis,* hypothetical contingency genes were found near chromosome ends (Kamper et al., 2006); however, their function remains unknown.

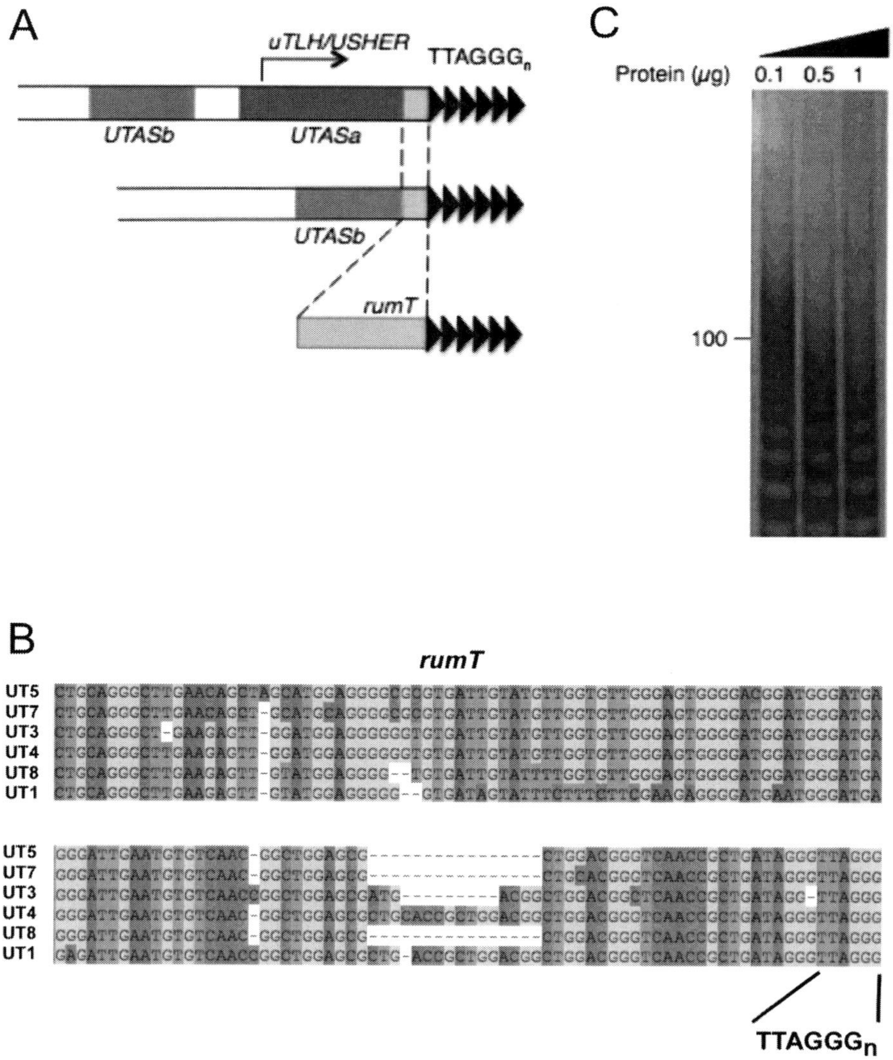

Figure 1. Organization of chromosome ends in *U. maydis,* and the determination of telomerase activity. A) A schematic representation of chromosome ends in *U. maydis* is depicted. Black arrowheads represent telomere repeats. Two main types of chromosome ends were found in the fungus: *UTASa*, which is a highly conserved element frequently found near telomere repeats, and *UTASb*, which is a less conserved element collinear to telomeres, and in some cases adjacent to *UTASa*. Conserved sequences, termed *rumT*, were found located between the telomere and TAS in all telomeres. B) Sequence alignment of *rumT* from six telomeres. A string of imperfectly repeated motifs was found at a variable region of the border whose meaning is currently unknown. C) Telomerase activity in *U. maydis*. Protein extracts from the FB1 strain was diluted to 0.1, 0.5, and 1.0 μg and analyzed for telomerase activity. As protein concentration increases, telomerase activity decreases, which is possibly related to PCR inhibitory contaminants. The template used for telomere lengthening was from the telomere border *rumT* (US primer). The results were confirmed using a commercial TRAP-ELISA kit (Roche), which uses a different primer in the assay. The length of the product is shown in base pairs on the left.

In the vicinity of the telomere, the expression of genes is subject to the telomere position effect (TPE), which drives variegated phenotypes derived of reversible silencing by the chromatin dynamics. TPE was first reported by Gottschling et al., (1990) and is defined as the exertion of reversible silencing on the genes located at the telomere boundary. It has been explained in several ways, ranging from remodeling of the higher order chromatin organization in response to physiological states of the cell, to cell cycle regulated expression of non-coding RNAs. In all cases, TPE underlies the recurring post-translational modification of chromatin. Epigenetic regulation of telomeres has been the subject of excellent reviews for organisms ranging from human to yeast (Schoeftner & Blasco, 2010; Tham & Zakian, 2003). In *U. maydis,* the exploration of *cis*-acting sequence-mediated silencing has begun recently. Preliminary experiments using autonomous replicating plasmids suggest that TPE can occur on a carboxin resistance cassette (Cbx^r) by telomeric repeats of *U. maydis* or even *Arabidopsis thaliana* when it is cloned at the 5' side of its promoter (Sanchez-Alonso, unpublished). A decrease in the transformation frequency by 1/50 was observed when the UT1 clone carrying $TTAGGG_{37}$ was cloned 200 bp upstream of the *Eco*RI/*Sst*I fragment that carried a mutant IP subunit from succinate dehydrogenase of pCBX122 that confers carboxin resistance (Keon et al., 1991). This result was unexpected, since it was hypothesized that plasmid stabilization would occur as reported by Longtine et al. (1992). When the non-replicating plasmid derivative pCB-A was used, the transformation frequency increased by 10-fold and plasmid stabilization increased by approximately 40%. Similar results were obtained when a spontaneous UT1 derivative (UTR3) carrying $TTAGGG_5$ was inserted at the same sites on pCBX122 and pCB-A derivatives. Our results assessing plasmid stabilization are in agreement with those obtained previously by Longtine (1992) regarding the improvement of circular plasmid segregation by telomere-repeat sequences in *S. cerevisiae*, and with those obtained by Javerzat (1993) and Takahashi (2011) showing enhancement of the autonomous-replicating properties and stabilization of linearized plasmids carrying telomere-like repeats in *Podospora anserina* and *Cryptococcus humicola*, respectively. However, these results disagree with the silencing of Cbx^r in our system. Plasmids harboring the silent mating locus HMR-E and presumably telomere repeats enhance its segregation efficiency by binding to silencing proteins Rap1, Sir2, Sir 3, and Sir 4. It is likely that these plasmids and natural telomeres co-localize as clusters in the nuclear periphery and function as segregating structures (Kimmerly & Rhine, 1987; Gehlen et al., 2011). The TPE on Cbx^r expression was dramatic, and the introduction of a single telomere repeat from *U. maydis* or even *A. thaliana* decreased the transformation frequency. In *S. cerevisiae,* telomere-associated sequences have been described as composites with an abundance of *cis*-acting sequences, such as the X-*ARS* consensus sequences (*X-ACS*) and Abf1-binding site, which behave as silencing relayers (proto-silencers) of the silencing that is exerted by the telomere repeat. Insulators, which were first described as subtelomeric anti-silencing regions (STARs) of yeast, counteract silencing and are represented at TAS by Tbf1 and Reb1 binding sites (Fourel et al., 2000; Pryde & Louis, 1999). Both protosilencers and STARs create discontinuous patterns of silencing at subtelomeric regions. Recently, it was found that insulators are members of *cis*-acting sequences, including enhancers and activators, which operate through interactions with transcription factors, including transcription factors responsive to the environment. In addition, the delicate equilibrium of insulators suppresses extreme variations in gene silencing at subtelomeric regions (Fourel et al., 2004; Power et al., 2011; Smith et al., 2011). We hypothesize that our findings were due to the location of Cbx^r beyond the telomere repeat in our system. Therefore, in the absence of any other intervening subtelomeric sequence, the silencing exerted by the cloned telomere affected the expression of the gene resistance but not the segregation efficiency. Further analysis may provide additional insight on telomere silencing in *U. maydis* by *cis*-acting elements.

THE TELOMERASE IN *U. MAYDIS*

Telomerase is a multimeric ribonucleoprotein complex with specialized reverse transcriptase activity that synthesizes telomere repeats from the G-rich strand in order to maintain telomere length homeostasis (Zvereva et al., 2010). The catalytic core of telomerase is composed of a specialized reverse transcriptase (TERT) and an RNA subunit (TR) that serves as a template for telomere synthesis. Mutations affecting the catalytic core genes have dramatic effects on telomere length, chromosome stability, and cell viability. A third essential component, dyskeryn (DKC), is a small-nucleolar ribonucleoprotein (snoRNP) with pseudouridine synthase activity that associates with the catalytic core during telomerase biogenesis (Cohen et al., 2007). Telomerase solves the end-replication problem that arises from the failure of conventional polymerases to replicate open chromosome-ends, which causes telomeres to shorten in each replication round (Chan & Blackburn 2004). In cells lacking telomerase activity, the telomeres shorten to a critical point known as M1; at this point, cellular proliferation capability ceases and senescence occurs (Masutomi et al., 2003). In higher eukaryotes, however, not all cell types posses significant quantities of telomerase activity to replenish the lost telomeric repeats, and only cell lineages committed to proliferate significantly possess sufficient levels of the enzyme. Telomerase activity has been detected in cancer cells lineages, germinal and pluripotential cells, and unicellular eukaryotes, which could be considered as immortal cell stocks (Kim et al., 1994; Flores et al., 2006). In metazoans, the expression and activity of telomerase seems to be strictly regulated throughout the cell cycle and is negatively regulated as cellular and tissue development programs progress in the life cycle of metazoan organisms (Armstrong et al., 2000; Kyo et al., 2008).

Telomerase has been the focus of intense research. Currently, two general functions of the TERT subunit of telomerase are recognized: the first and well-studied telomere maintenance function, and the second and more recently uncovered series of telomere-independent functions. In addition to the necessity of dyskerin hnRNP for telomerase biogenesis, telomere synthesis requires the TERT subunit and RNA template, and these components are limiting factors of telomerase (Flores et al., 2006; Wong et al., 2006). Since the catalytic subunit is the only necessary factor for restoration of telomerase activity in many organisms, such as human and mouse, most of the studies on telomerase function initially focused on the regulation of TERT expression (Bodnar et al., 1998; Kyo et al., 2008). As an enzyme complex, telomerase is exquisitely regulated: the hTERT promoter is rich in binding sites for transcriptional factors, such as Sp1, Ap1, and c-Myc, and therefore its transcription is strictly regulated during the developmental programs of the cell. Other control mechanisms include mRNA alternative splicing, multimerization and phosphorylation of the telomerase catalytic subunit, and the proper assembly of the higher-order telomeric DNA-protein complexes (Ulaner et al., 2000; Arai *et al*., 2002; Kang et al., 1999; Li et al., 1997; Gonzalo et al., 2005). In addition, a series of more recent observations in various model organisms, including mouse, human, and Xenopus, have shown that TERT is involved in extra-telomeric functions that are critically important for cell development (Ahmed et al., 2008 and references therein). These telomere-independent functions include: (a) serving as a cofactor of BRG1 to modulate the WNT-β-catenin signaling pathway, which activates Wnt responsive genes that are important for stem cell proliferation in some vertebrates (Park et al., 2009), (b) directly interacting with the RNA component of mitochondrial RNA processing (RMRP) endoribonuclease to produce dsRNAs that are processed as siRNA by the Dicer protein (Maida et al., 2009), and (c) co-localizing with mitochondria to protect cells from oxidative stress (Sahin et al., 2011). Importantly, it was found that telomere dysfunction impairs mitochondria biogenesis.

Identification of the telomerase reverse transcriptase (*utert1*) gene and the characterization of telomerase activity in *U. maydis* have also been more recently described. Using the TRAP protocol with either the *Telo*TAGGG telomerase PCR ELISA Kit or the TRAP system as described by Kim et al., (1994), we consistently observed telomere synthesizing activity in cell extracts of *U. maydis,* but at lower levels than those obtained from the human embryonic cell line 293. A TRAP assay using cell extracts showed the influence of protein quantities on the TRAP assay using the telomere border from *U. maydis* chromosomes as a substrate for telomere elongation *in vitro* (Figure 1C). Although the TTAGGG telomere repeated motif builds human and *U. maydis* telomeres, differences in telomerase activity were found in both experimental systems. It is possible that the results could be attributed to protocol adaptation, however, the natural quantity and quality in the intrinsic activity of telomerase enzyme in each organism should be explored, since differences in the structure of telomerase from higher and lower eukaryotes have been previously reported (Hug & Ligner, 2006) as well as in the chromatin structure of telomeres from yeast and humans (Blasco, 2007). These differences could substantially contribute to substrate affinity from those distantly related telomerases under the experimental conditions used, and it would be interesting to determine if the telomere length attained after the elongation step also contributes to the activity of telomerase *in vitro.*

The TERT gene was predicted by pairwise alignment of the *Homo sapiens* TERT gene (NP_937983.2) with sequences of *U. maydis* genome (http://www.broad.mit.edu) as the product encoded by the first ORF from the locus UM00761.1 (score 209.5, E = 0.0). (Sanchez-Alonso, unpublished). The predicted protein shared all of the characteristic motifs of the RT domain, RNA-binding domain, and N-terminal domain as other well-characterized telomerases. Although the function of each of these proposed motifs will require further biochemical characterization, the distances between the motifs matches the ratios reported for other TERT genes (Kelleher et al., 2002). Disruption of the *utert1* gene induced characteristic telomere shortening in the telomerase-deficient mutants as well as replicative senescence, which was manifested as stressed, slowly growing, and poorly replicating colonies. This phenotype was similar to that observed in *S. cerevisiae* telomerase-defective mutants lacking *TLC1* gene, which encodes the RNA subunit of the yeast telomerase (Lundblad & Szostak, 1989). Post-senescent survivors also arose from telomerase-deficient mutants in *U. maydis* (Sanchez-Alonso, unpublished). In *S. cerevisiae* mutants lacking telomerase activity, two general types of survivors were reported on the basis of the amplification pattern obtained from sequences of the chromosomal termini: Type I, which showed a recombination-based amplification of subtelomeric Y' elements and required *RAD51* and *RAD52* functions, and Type II, in which long and heterogeneous telomere repeats were observed as amplified products that required *RAD50, RAD52*, and *SGS1* functions (Teng & Zakian, 1999; Huang et al., 2001). In *U. maydis,* studies on the survivor classes similar to those of yeast and the possible recombinational mechanisms involved in these subtelomere events are currently under investigation.

The relationship between replicative potential and telomere lengthening has also been studied in various higher eukaryotes (Blagoev, 2009; Grafi et al., 2007). hTERT and the epigenetic status of telomere and subtelomeric sequences are important for growth, proliferation and differentiation of human embryonic or induced stem cells (Yang et al., 2008; Yehezkel et al., 2011). Moreover, the sole reactivation of telomerase activity restores the replicative potential and neurological functions of aged, telomerase-deficient mice (Jaskelioff et al., 2011). With regard to the role of *utert1* in life cycle completion of *U. maydis,* we found *utert1* important for teliospore formation, either through its involvement in telomere-related or telomere-independent functions in the differentiation of fungal cells. Mechanisms associated with teliospore formation in telomerase-negative cell are currently being studied in our group, since the replicative potential and viability of *U. maydis* could have a role in successful invasion of the plant. That hypothesis is in agreement with the findings of Walbot & Skibbe (2010), which showed that host growth potential is one factor that influences the pathogenic potential of *U. maydis*. Additional research on the functions of telomerase in pathogen microorganisms could provide additional insight regarding the mechanisms involved in disease persistence.

Chromosome Ends in Wild *U. Maydis* Isolates and Laboratory Strains

Although the overall organization of sequence elements from the subtelomeric region of *U. maydis* resembles that of *S. cerevivise*, clear differences between Y' elements and *UTASa* were found in sequence homology and length. Y' elements harbor two overlapping open reading frames that lack homology to any reported proteins, with the exception of a short stretch that has local identity with RNA helicases in both cases. One of these ORFs encodes a DEXH subfamily member that has been known until now as Y' *Hel*icase *p*rotein *1* (Y'-Help1), which shares homology with initiation factor 4A from eukaryotes and with RecG of *Aquifex sp.* This element is barely affected by sequence variations occurring in that subtelomeric regions, which affect the other ORF (Louis & Haber, 1992; Yamada et al., 1998). Y'-Help1 is poorly transcribed under ordinary growth conditions, but the gene is up-regulated in telomerase-negative survivors and transcripts can be easily detected. This finding has led to the hypothesis that Y'-Help1 participates in telomere metabolism; however, its function has not been well defined as yet (Yamada et al., 1998). In *U. maydis,* our laboratory was the first to describe homologous sequences to helicases in subtelomeric regions of a filamentous (dimorphic) fungus (Sanchez-Alonso & Guzman 1998). As a Y' element, *UTASa* seems to be a highly polymorphic sequence in both copy number and location on subtelomeric regions of several analyzed strains; however, its organization on chromosome ends and distribution amongst strains from different geographic origins is just beginning to be explored.

Table 1. Telomere-linked helicases (TLH) in fungi

Organism	Accession	Maximum length (aa) reported	DEXH box & (zinc binding domain)	Chrom. end location	Reference
Ustilago maydis	AAB95265	863	DECH (none)	Yes	Sánchez and Guzmán, 2002. Direct submission
Ustilago maydis (UmBLM)	XP_759021	1291	DEAH (RQC)	No	Birren et al. 2006 Direct submission
Ustilago maydis (RQL1)	XP_760820	748	DEAH (none)	No	Birren et al. 2006 Direct submission
Aspergillus fumigatus	EDP48545	706	DECH (none)	Yes	Fedorove et al., 2007 Direct submission
Metarhizium anisopliae	EFZ03297	2187	DECH (C2H2)	Yes	Gao et al., 2011 Direct submission
Talaromyces stipitatus	XP_002482995	1373	DECH (none)	Yes	Nierman et al., 2007 Direct submission
Pyrenophora tritici-repentis	XP_001942465	1422	DECH (C2H2)	Yes	Birren et al., 2007 Direct submission
Magnaporthe grisea	AAL86613.1	818	DECH (none)	Yes	Gao et al., 2002 Genetics 162,103
Magnaporthe orizae	ACP42042.1	1545	DECH (C2H2)	Yes	Rehmenyer et al., 2009 Curr. Genet. 55, 253
Penicillium marneffei	XP_002143128	1630	DECH (C2C2)	Yes	Fedorova et al., 2008 Direct submission
Schizosaccharomyces pombe	Q5EAK4	2100	DEAH (TDT)	Yes	Mandell et al. 2005 J. Biol. Chem 280, 5249
Pleurotus sp.'Florida'	CAR47902	880	DEAH (none)	Yes	Perez et al. 2009 Appl. Environ. Microbiol. 75, 1427
Saccharomyces cerevisiae (Sgs1)	P35187	1447	DEAH (RQC)	No	Gangloff et al. 1994 Mol. Cell. Biol. 14, 8391
Homo sapiens (WS)	Q14191	1432	DEAH (RQC)	No	Yu et al. 1996 Science 272, 258

TLHs from several fungal species were analyzed *in silico*; Because TLHs are polymorphic, the maximum length of those sequences annotated in databases is recorded in the third column; helicase motif II sequence, and zinc binding domain (ZBD) found in long TLH, and in RQC domain in RecQ orthologs. TDT refers to tellurite resistance/dicarboxylate transporter domain, which was found in the telomere linked helicase *tlh1* from *S. pombe*. The last column on the right refers to the authorship.

*Us*tilago *he*licase *r*ecQ-like (here named *USHER*) is the putative TLH from *U.maydis* and exhibits homology with several other telomere-linked helicases that have been reported in different organisms. Those sequences were annotated in public sequence databases as a result of genomic sequencing projects or recently reported (for references see Table 1). The sequences share the core helicase domain from RecQ enzymes with all seven conserved motifs from this subfamily, the conserved DECH sequence motif II, and in many cases a zinc binding domain (ZBD) at the N-terminal region (Gorbarenya et al., 1989; Vindigni & Hickson, 2009). The helicase encoded by *UTASa, USHER*, shares all seven conserved motifs of RecQ helicases, but lacks the HRDC domain as well as the well conserved RQC domain present in most RecQ helicases downstream of the helicase domain. The RCQ domain in turn is divided into two subdomains, a Zn^{+2} binding domain (ZBD) and winged helix domain, which are absent in TLHs at the canonical position for RecQ helicases (Vindigni & Hickson, 2009). Instead, the ZBD is present at the N-terminal region of putative helicases when they are larger than 1400 amino acids in length. This raises the possibility of "large" and "short" versions of TLHs occurring in subtelomeric regions, which exist in Y' elements of *S.*

cerevisiae, and opens an assortment of doubts about how many functions they could have. Sequence analysis by pairwise comparison shows a higher homology between *USHER* and *TAH1* from *Metarhizium anisopliae* (max. score 525, E_ 10^{-160}; Gao et al., 2011). Although sequences homologous to *UTASa* and *UTASb* were reported as telomere boundary elements at chromosome ends in *U. maydis*, the existence of other sequences on telomeric regions of *U. maydis* that resemble the N-terminal domain from the putative telomere-associated helicase of *M. anisopliae* have also been reported, which surprisingly conserve the ZBD (Kämper et al., 2006). Based on these findings, it is possible that *UTASa* is a subtelomeric element of equal length as *TAH1* from *M. anisopliae*, as was predicted by Rehmeyer et al., (2009). In fact, the sum of the sequence length of *USHER* and of those similar to *TAH1* match the average length found for larger TLHs. Laboratory strains UCM521, UCM520, I2, FB1, and FB2 of *U. maydis* was analyzed either by Southern blot or by PCR analysis for the occurrence of *USHER* at chromosome ends (Figure 2). A strong hybridization signal to *USHER* was found in all laboratory strains with a similar distribution pattern (Figure 2A). In the I2 strain, the signal was weak, which was expected because the I2 strain was previously shown to have a reduced amount of *UTASa* sequences on the chromosome ends (Sanchez-Alonso & Guzman, 1998). In addition, a PCR analysis was designed to amplify sequences from the predicted catalytic core and adjacent sequences (~600 bp) and then digested with the *Sst*I endonuclease enzyme to verify the expected sequence (Figure 2B).

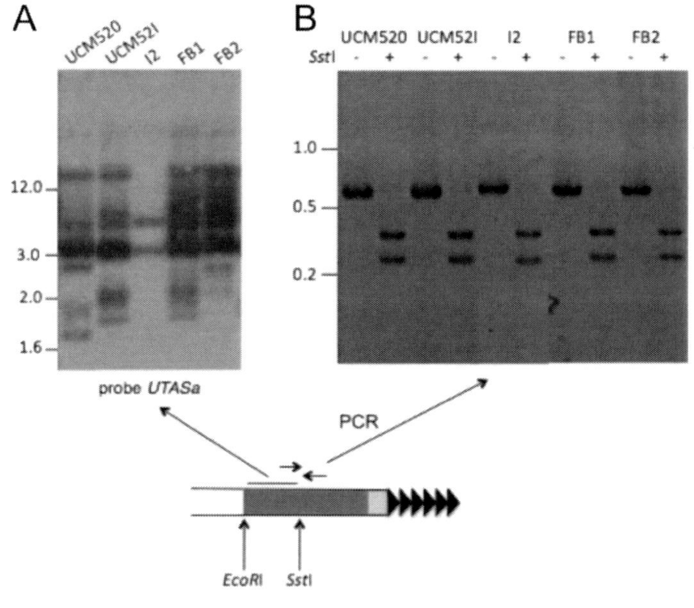

Figure 2. Telomere-linked helicases (TLH) sequences in *U. maydis* laboratory strains. A) Analysis by Southern blot. Total DNA from specified strains in the was digested to completion with *Eco*RI restriction endonuclease, and the fragments were resolved by electrophoresis on a 0.8% agarose gel and then transferred to nylon (Immobilon, Millipore). Hybridization was carried out using the reported UT4-a (Sanchez-Alonso and Guzman 1998). B) Analysis by PCR amplification. Amplification of *UTASa* was carried out with forward (AVV1) and reverse (AVV2) oligonucleotide primers (represented by arrows), that span helicase motifs IV, V, and VI, which preserve the *Sst*I restriction site. Strain names are shown above the amplified DNA. The amplified DNA fragments (-) were digested with *Sst*I (+).

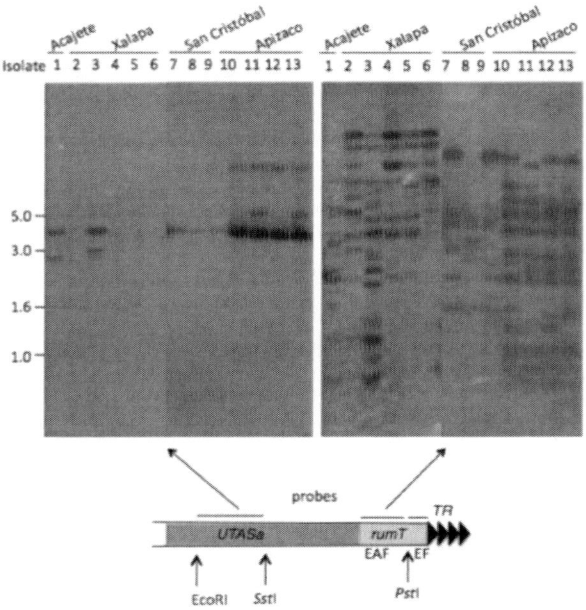

Figure 3. TLH sequences in wild isolates of *U. maydis*. The prevalence of *USHER* in wild-type isolates of *U.* maydis was analyzed using 150 wild isolates of *U. maydis* collected from four geographical locations in Puebla, Veracruz, and Tlaxcala states of Mexico. Hybridization was carried out using the previously reported UT4-a probe (Sanchez-Alonso and Guzman 1998) for *UTASa* detection. Despite the lack of UT4-a hybridization signal from Xalapa strains, amplification analysis of the *USHER* helicase domain was successful by PCR. No strains were identified without *UTASa* sequences among the 150 isolates.

The occurrence of *UTASa* on chromosomes from different *U. maydis* strains appears to be as variable in sequence as in copy number. However, *UTASa* is not structurally or functionally essential at individual telomeric regions, and therefore this element may be necessary for DNA metabolism under conditions not yet explored. At the transcript level, subtelomeric sequences could be involved in the control of chromatin condensation at subtelomeric regions or participate in TPE events. As a protein product, the importance of this sequence will require further elucidation, since the expression of a hundred copies of a helicase gene could have catastrophic effects on cells. However, under controlled expression, the sequence could still have a role during senescent stages. In *Magnaporthe oryzae,* only four full-length copies of the TLH element were found at the chromosomal ends, with the remaining copies having truncations, missense mutations, or interruptions (Rehmeyer et al., 2009). Under this type of scenario, it could be possible that the telomere dynamics are a force that limits over-expression of those telomeric elements. No RecQ activity has been reported for TLHs to date, however, in *Schizosaccharomyces pombe,* ectopic expression of a subtelomeric RecQ helicase homolog enables telomerase-deficient cells to recover from crisis earlier than those expressing a mutant isoform of the same helicase, and the authors have proposed that TLH is a second RecQ helicase (Mandell et al., 2005). In *S. cerevisiae,* Y'-Help1 exhibits helicase activity when cloned and expressed from the YES2 expression vector, but no function was attributable to that activity (Yamada et al., 1998). *UTASa* is an unfailing element from the telomeric region of *U. maydis* on some, many or most of the chromosomes of *U. maydis* (data not shown). The sequence does not have prevalence on any specific chromosome, but the size, distribution, and function has not been determined. In addition, no strains have been found that lack *UTASa* sequences among 150 isolates that were analyzed by Southern blot or PCR assay. A representative Southern analysis showing the prevalence of *USHER* in several wild isolates of *U. maydis* is shown in Figure 3. Isolates were collected from four geographical locations in Puebla, Veracruz, and Tlaxcala states of Mexico. The reported UT4-a was used to identify a *UTASa* hybridization pattern amongst different isolates (Sanchez-Alonso & Guzman 1998). Despite the lack of a UT4-a hybridization signal from Xalapa strains, the amplification analysis of the helicase domain was successful.

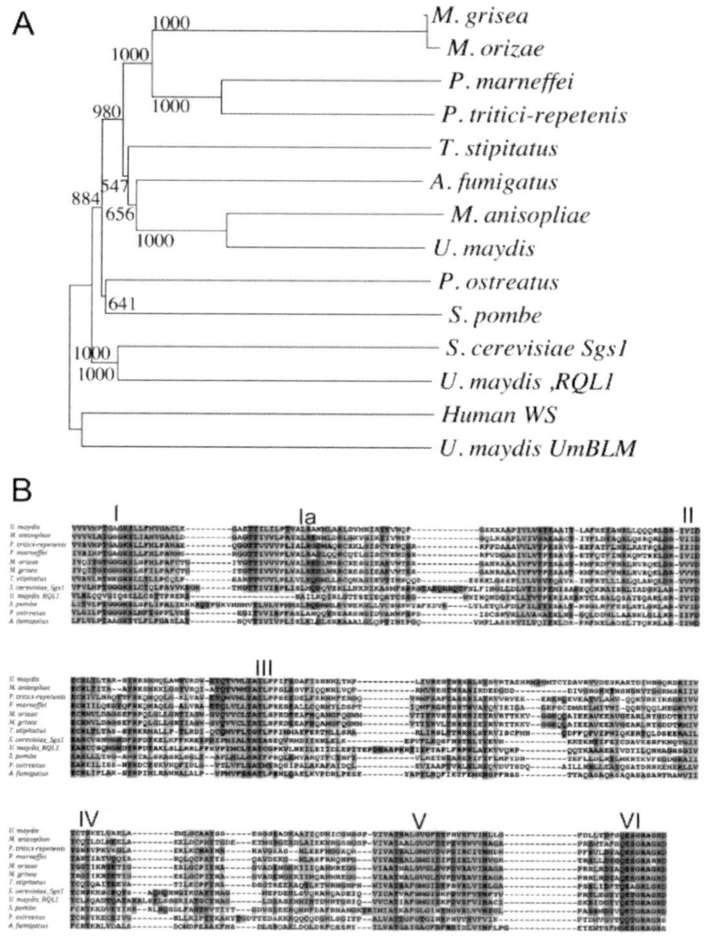

Figure 4. Phylogenetic relationships of TLHs in fungi. Multiple alignment sequence analysis of both characterized and hypothetical telomere-linked helicases (TLH) was conducted in diverse fungal species. A) The phylogenetic tree of TLH was constructed using sequences spanning the helicase domain from the conceptual helicase proteins (accession numbers in Table 1). The tree was created using the neighbor-joining method and was performed to determine phylogenetic relationships. B) Protein sequence alignment of the predicted TLH, including that of *U. maydis* and RecQ ortholoquess, RecQ helicase domain used in this analysis includes the seven conserved motifs shared by RecQ ortholoquess and TLHs. Differences in motif II were found in members of the two groups.

TELOMERE-LINKED HELICASES (TLH) IN FUNGI

UTASa shares homology with nine other putative TLHs from the chromosome ends of diverse fungal species. This observation is remarkable, since this well conserved element was found on chromosome ends from distantly related organisms, such as basidiomycetes and ascomycetes. A compilation of TLHs from different fungi is listed in Table 1, and Y'-Help1 was excluded because it is not RecQ related. TLHs were analyzed for their structure accordingly to the findings by Rehmeyer (2009), which included total length, DEXH motif, and a novel C2H2 or ZBD in the N-terminus side of the conceptual products. From those putative helicases, nine were identified in ascomycetes and two were identified in basidiomycetes, including *USHER*. On the basis of a size-length analysis and as well as reports on the scarce number of full-sized TLHs compared to truncated versions on chromosome ends (Mandell et al., 2005; Rehmeyer et al., 2009), it is conceivable to propose larger versions of TLHs that have a ZBD in the remaining fungi analyzed. However, one must be cautious to make generalizations concerning subtelomeric sequences due to its variable nature. Previous speculation has been made regarding the roles of these putative helicases in telomere metabolism, conferring advantages to fungi for survival under stress, improvements in adaptation to novel environments, or as a component of general DNA metabolism (Mandell et al., 2005; Gao et al., 2002; Rehmeyer et al., 2009). TLHs have been found distributed on the first 10 kb of the chromosome, and most of them have been oriented 5' to 3' toward the chromosome ends. It is possible that novel mechanisms involved in TLH protection are also operating to preserve full length elements, such as in *Drosophila sp.,* whereas truncated versions has as yet unknown functions (George et al., 2010).

Phylogenetic analysis has been the keystone in studies of the non-coding sequences adjacent to telomeres. TLHs were found in ascomycetes and basidiomycetes to share related global structure and organization at chromosome ends. Using only the sequences of the RecQ-helicase domain from the compiled helicases, a neighbor-joining tree was constructed by comparing selected sequences from the catalytic domain of *USHER* with those of the remaining eight subtelomeric helicases (Figure 4A). The tree contains two clades, each of which includes members of both ascomycetes and basidiomyces taxonomic groups. A minor clade included *S. pombe* and *Pleurotus sp.* "Florida", and the major clade was built from ascomicetous fungi and *U. maydis*. In addition, both clades of TLHs were separate from the yeast and human RecQ orthologues Sgs1p and Werner Syndrome protein, respectively. The results, which were well supported by bootstrapping with 1000 replicates, argues for an ancestral relationship and strengthens the proposal of an indispensable function exerted by TLHs when conserved by any particular species. It is unknown if all filamentous fungi harbor TLH sequences or if these sequences are exclusive of some species and strains, since it is hard to take a deep approach for the characterization of telomere-associated sequences in genome sequencing projects due to the repetitive nature and interspersed organization. It is also unknown whether TLHs only occur in specific strains, such as *M. grisea*, whose isolates from non-rice hosts seem to lack TLH based on the results obtained by Southern blot analysis of diverse isolates (Gao et al., 2002). In the case of *U. maydis,* PCR analysis has shown that all tested strains conserve at least the RecQ domain of *USHER*, even though in those in which the hybridization signal is substantially weak for detection by autoradiography. A multiple alignment of the core sequences of TLH from the helicase domain was performed, which spans the seven conserved motifs of the RecQ helicase subfamily (Figure 4B). The RecQ orthologues Sgs1p and Werner protein from yeast and human, respectively, were also included as roots. The alignment was performed with the Clustal X algorithm using the sequences listed in Table 1. These results support the hypothesis of different functions for both groups (RecQ and TLHs) on the basis of the DEXH motif, despite the idea that these helicase groups may play a role in telomere maintenance.

Detection of TLH Transcripts in *U. Maydis*

The fungal telomere and telomere-associated sequences assemble as nucleoprotein complexes that share properties with heterochromatin of higher eukaryotes. Genes harbored in these regions are subjected to transcriptional repression called TPE, as described above. In yeast, the telomeres act as the master silencer that emanates the resilient repressive signals that are responsible for TPE, whereas TAS, mating type *loci HML* and *HMR*, as well as *rRNA* spread the transcriptional repression to remote distances from telomeres and act as relayers or proto-silencers (Fourel et al., 2004). Silencing requires *cis*-acting elements, such as ARS consensus sequences (ACS) and Rap1p and Abf1 binding sites. In addition, this mechanism requires protein factors ORC, Rap1p, Abf1, and the Sir2/3/4 protein complex as well as other more recently reported environment-responsive transcription factors (Louis et al., 1994; Smith et al., 2010). Sir2, which is a histone deacetylase in the complex, is a key enzyme for silencing management; Sir2, together with Sir3 and Sir4, bind as a complex to Rap1p, the most abundant telomere-binding protein in yeast, and also recruits Rif1/Rif2 proteins in a Y'- or X-dependent manner (Iglesias et al., 2011). Sir2 deacetylates histone H4 to initiate chromatin condensation, whereas histone acetyltransferase Sas2p counteracts H4 deacetylation to de-condensate chromatin (Tham & Zakian, 2002; Kozak et al., 2010). A subtle equilibrium between aceylase/deacetylase activities establishes a "negotiable border", where limits are dynamically regulated in length and characterize the TPE at telomere boundaries (Kimura & Horikoshi, 2004). Under these harsh conditions, transcriptional activity at the telomere boundary emerges mainly as a product of RNApol II and the polyadenylation by Pap1 (Luke et al., 2008; Shoeftner & Blasco, 2008). The first detected transcripts, called *T*elomere *R*epeat-containing RNA (TERRA), were reported in mammals and yeast but are presumed to be also present in distinct genera from different taxonomic groups (Azzalin et al., 2007; Schoeftner & Blasco, 2008). In initial assays, transcripts were detected by RT-PCR from samples of total RNA extracted from non-synchronous cell cultures of different strains of *U. maydis* (Figure 5). The *USHER* transcripts were screened for either sense (reverse primer, top of Figure 5) or antisense orientation (forward primer, top of Figure 5). Amplified segments from cloned *USHER* were used as hybridization probes, and positive signals were obtained for both sense and antisense amplified cDNA (Figure 5, bottom panel). Semi-quantitative PCR amplification of cDNA synthesized from RNA obtained from synchronous cultures of *U. maydis* confirmed the variations in *USHER* transcript abundance during cell cycle. As shown in Figure 6, which shows two replication cycles, a high *USHER* transcript level was observed at the initiation of each cycle. Similar experiments using oligo dT_{12-18} oligonucleotides or Kim's CX oligonucleotide (3'-A TC CCATTC CCATTC CCATTC-5') to synthesize the first strand suggest that TERRA is present in *U. maydis* (Kim et al., 1994). Analyses to determine whether telomere repeats are present in those transcripts and to confirm their identity as long non-coding RNA from TAS or as TERRA are ongoing. Moreover, Usher protein was absent or undetectable in wild-type strains when polyclonal antibodies where used to detect it, and identification of Usher in protein extracts from telomerase-negative mutants would provide additional information about *USHER* metabolism and function in this fungus.

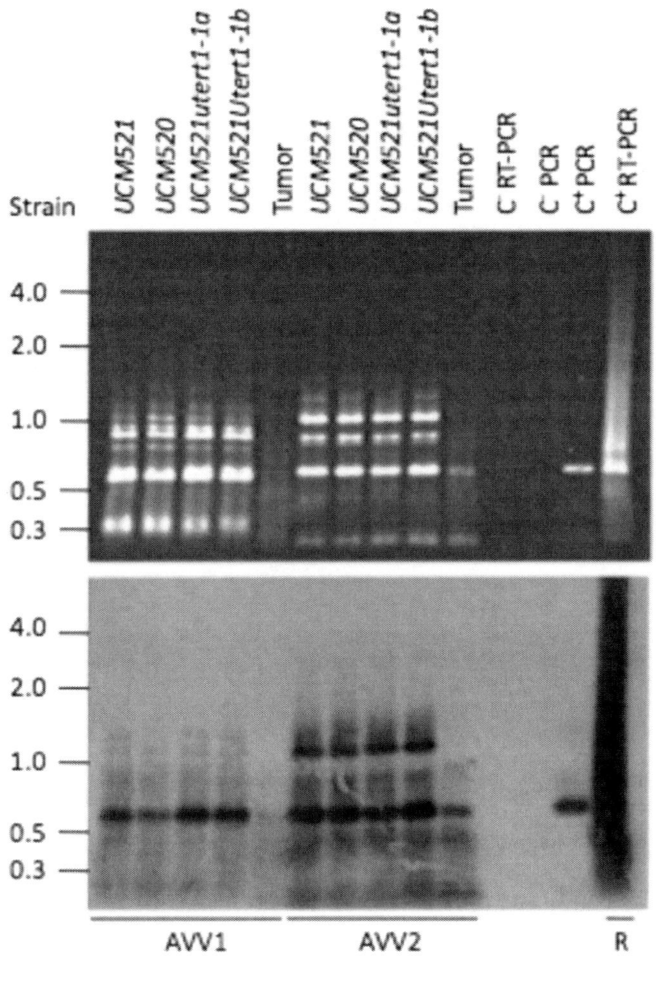

Figure 5. Detection of sense and antisense TLH transcripts. Total RNA (DNA free) from laboratory strains UCM501, UCM520, UCM521, and two utert-strains (UCM521utert1-1a at 54 and UCM521utert1-1b at 70 doubling times after disruption) as well as from UCM520 X UCM521 (Tumor) was used for reverse transcription with forward (AVV1) or reverse (AVV2) primers; see the location of the primers in Figure 2. An aliquot of each one of the cDNA first strands was then PCR amplified with both primers. The top panel, shows the amplified DNA resolved by agarose gel electrophoresis and the botton panel the Southern blot hybridized with the *UTASa* probe (see Figure 2); a similar hybridization signal at 600 bp is detected from each one of the reverse transcription assays. No reverse transcription and no PCR controls are shown, C^-RT-PCR and C^-PCR respectively. PCR products from cloned USHER and RT-PCR products using total RNA from an utert1- survivor are shown as positive controls, C^+PCR and C^+RT-PCR, respectively.

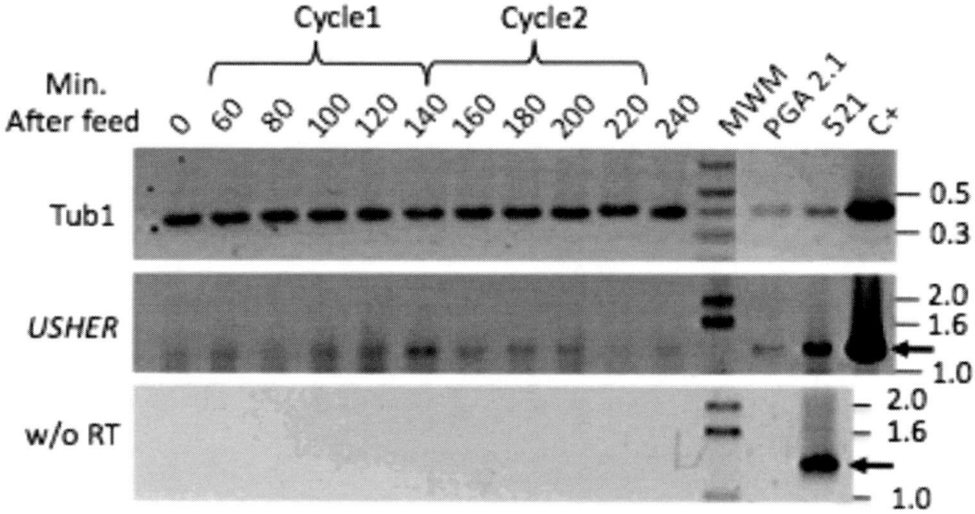

Figure 6. TLH transcript variation during the cell cycle. Semi-quantitative analysis using RT-PCR was carried out with RNA extracted from samples removed every 20 min from synchronous cultures of *U. maydis*. Two replication cycles confirm the variations in *USHER* transcript abundance during the cell cycle, each of which initiates with a high level of transcript. PGA1 and UCM521 were used as PCR controls for high and low copy numbers of *USHER*. w/o RT, RNA without reverse transcription.

Conclusion

The determination of telomere structure and organization in *U. maydis* as well as its telomerase characterization is the latest and still poorly studied subject of telomere metabolism. However, the results that have been obtained to date strongly agree with those obtained in *S. cerevisiae*. The determination of telomerase activity using TRAP analysis is an amenable strategy to use in chromosome metabolism studies in this fungus and provides clear and reproducible results when applied to sporidia cell cultures. Measurements of telomere size and length by Southern blot throughout the TRF analysis was previously conducted and shown to be in a satisfactory range and composition for studies on telomerase/telomere maintenance and dynamic equilibrium. Similar to other model systems, this fungus also shares the machinery that allows for the emergence of survivors from telomerase-deficient mutants as well as recombination systems that operate as alternative pathways to maintain telomere function. On the other hand, *U. maydis* also displays special advantages of morphologic transitions related to its developmental stages during its life cycle. As a plant pathogen, *U. maydis* is amenable to the exploration of the phenotypic consequences of different expression levels of telomerase components during its life cycle, the consequences of deficiencies in these components on proliferation and pathogenic potential, and the study of telomerase regulation. This article has provided a general overview of the studies conducted to date in a interesting and alternative model organism, *U. maydis,* but more detailed studies are necessary to precisely define the contributions that this fungus can offer to our understanding of telomeres and telomerase.

REFERENCES

Ahmed, S., Passos, J. F., Birket, M. J., Beckmann, T., Brings, S., Peters, H., Birch-Machin, M. A., von Zglinicki, T. & Saretzki, G. (2008). Telomerase does not counteract telomere shortening but protects mitochondrial function under oxidative stress. *Journal of cell science, 121*, 1046-1053.

Arai, K., Masutomi, K., Khurts, S., Kaneko, S., Kobayashi, K. & Murakami, S. (2002). Two independent regions of human telomerase reverse transcriptase are important for its oligomerization and telomerase activity. *J Biol Chem, 277*, 8538-8544.

Armstrong, L., Lako, M., Lincoln, J., Cairns, P. M. & Hole, N. (2000). mTert expression correlates with telomerase activity during the differentiation of murine embryonic stem cells. *Mech Dev, 97*, 109-116.

Azzalin, C. M. & Lingner, J. (2008). Telomeres: the silence is broken. *Cell Cycle, 7*, 1161-1165.

Azzalin, C. M., Reichenbach, P., Khoriauli, L., Giulotto, E. & Lingner, J. (2007). Telomeric repeat containing RNA and RNA surveillance factors at mammalian chromosome ends. *Science, 318*, 798-801.

Barry, J. D., Ginger, M. L., Burton, P. & McCulloch, R. (2003). Why are parasite contingency genes often associated with telomeres? *Int J Parasitol, 33*, 29-45.

Behm-Ansmant, I. & Izaurralde, E. (2006). Quality control of gene expression: a stepwise assembly pathway for the surveillance complex that triggers nonsense-mediated mRNA decay. *Genes Dev, 20*, 391-398.

Biessmann, H. & Mason, J. M. (1994). Telomeric repeat sequences. *Chromosoma, 103*, 154-161.

Blagoev, K. B. (2009). Cell proliferation in the presence of telomerase. *PLoS One, 4*, e4622.

Blasco, M. A. (2007). The epigenetic regulation of mammalian telomeres. *Nat Rev Genet, 8*, 299-309.

Bodnar, A. G., Ouellette, M., Frolkis, M., Holt, S. E., Chiu, C. P., Morin, G. B., Harley, C. B., Shay, J. W., Lichtsteiner, S. & Wright, W. E. (1998). Extension of life-span by introduction of telomerase into normal human cells. *Science, 279*, 349-352.

Bourns, B. D., Alexander, M. K., Smith, A. M. & Zakian, V. A. (1998). Sir proteins, Rif proteins, and Cdc13p bind Saccharomyces telomeres in vivo. *Mol Cell Biol, 18*, 5600-5608.

Cesare, A. J., Groff-Vindman, C., Compton, S. A., McEachern, M. J. & Griffith, J. D. (2008). Telomere loops and homologous recombination-dependent telomeric circles in a Kluyveromyces lactis telomere mutant strain. *Mol Cell Biol, 28*, 20-29.

Chan, C. S. & Tye, B. K. (1983). Organization of DNA sequences and replication origins at yeast telomeres. *Cell, 33*, 563-573.

Chan, S. R. & Blackburn, E. H. (2004). Telomeres and telomerase. *Philos Trans R Soc Lond B Biol Sci, 359*, 109-121.

Cohen, S. B., Graham, M. E., Lovrecz, G. O., Bache, N., Robinson, P. J. & Reddel, R. R. (2007). Protein composition of catalytically active human telomerase from immortal cells. *Science, 315*, 1850-1853.

Dai, X., Huang, C., Bhusari, A., Sampathi, S., Schubert, K. & Chai, W. (2010). Molecular steps of G-overhang generation at human telomeres and its function in chromosome end protection. *Embo J, 29*, 2788-2801.

de Bruin, D., Zaman, Z., Liberatore, R. A. & Ptashne, M. (2001). Telomere looping permits gene activation by a downstream UAS in yeast. *Nature, 409*, 109-113.

de Lange, T. (2010). Telomere biology and DNA repair: enemies with benefits. *FEBS Lett, 584*, 3673-3674.

Drinnenberg, I. A., Weinberg, D. E., Xie, K. T., Mower, J. P., Wolfe, K. H., Fink, G. R. & Bartel, D. P. (2009). RNAi in budding yeast. *Science, 326*, 544-550.

Feuerhahn, S., Iglesias, N., Panza, A., Porro, A. & Lingner, J. (2010). TERRA biogenesis, turnover and implications for function. *FEBS Lett, 584*, 3812-3818.

Flores, I., Benetti, R. & Blasco, M. A. (2006). Telomerase regulation and stem cell behaviour. *Curr Opin Cell Biol, 18*, 254-260.

Fourel, G., Magdinier, F. & Gilson, E. (2004). Insulator dynamics and the setting of chromatin domains. *Bioessays, 26*, 523-532.

Fourel, G., Revardel, E., Koering, C. E. & Gilson, E. (1999). Cohabitation of insulators and silencing elements in yeast subtelomeric regions. *Embo J, 18*, 2522-2537.

Gangloff, S., McDonald, J. P., Bendixen, C., Arthur, L. & Rothstein, R. (1994). The yeast type I topoisomerase Top3 interacts with Sgs1, a DNA helicase homolog: a potential eukaryotic reverse gyrase. *Molecular and cellular biology, 14*, 8391-8398.

Gao, Q., Jin, K., Ying, S. H., Zhang, Y., Xiao, G., Shang, Y., Duan, Z., Hu, X., Xie, X. Q., Zhou, G., Peng, G., Luo, Z., Huang, W., Wang, B., Fang, W., Wang, S., Zhong, Y., Ma, L. J., St Leger, R. J., Zhao, G. P., Pei, Y., Feng, M. G., Xia, Y. & Wang, C. (2011). Genome sequencing and comparative transcriptomics of the model entomopathogenic fungi Metarhizium anisopliae and M. acridum. *PLoS Genet, 7*, e1001264.

Gao, W., Khang, C. H., Park, S. Y., Lee, Y. H. & Kang, S. (2002). Evolution and organization of a highly dynamic, subtelomeric helicase gene family in the rice blast fungus Magnaporthe grisea. *Genetics, 162*, 103-112.

Gehlen, L. R., Nagai, S., Shimada, K., Meister, P., Taddei, A. & Gasser, S. M. (2011). Nuclear geometry and rapid mitosis ensure asymmetric episome segregation in yeast. *Curr Biol, 21*, 25-33.

George, J. A., Traverse, K. L., DeBaryshe, P. G., Kelley, K. J. & Pardue, M. L. (2010). Evolution of diverse mechanisms for protecting chromosome ends by Drosophila TART telomere retrotransposons. *Proc Natl Acad Sci U S A, 107*, 21052-21057.

Giraud-Panis, M. J., Teixeira, M. T., Geli, V. & Gilson, E. (2010). CST meets shelterin to keep telomeres in check. *Mol Cell, 39*, 665-676.

Gonzalo, S. & Blasco, M. A. (2005). Role of Rb family in the epigenetic definition of chromatin. *Cell Cycle, 4*, 752-755.

Gorbalenya, A. E., Koonin, E. V., Donchenko, A. P. & Blinov, V. M. (1989). Two related superfamilies of putative helicases involved in replication, recombination, repair and expression of DNA and RNA genomes. *Nucleic Acids Res, 17*, 4713-4730.

Gottschling, D. E., Aparicio, O. M., Billington, B. L. & Zakian, V. A. (1990). Position effect at S. cerevisiae telomeres: reversible repression of Pol II transcription. *Cell, 63*, 751-762.

Grafi, G., Ben-Meir, H., Avivi, Y., Moshe, M., Dahan, Y. & Zemach, A. (2007). Histone methylation controls telomerase-independent telomere lengthening in cells undergoing dedifferentiation. *Dev Biol, 306*, 838-846.

Griffith, J. D., Comeau, L., Rosenfield, S., Stansel, R. M., Bianchi, A., Moss, H. & de Lange, T. (1999). Mammalian telomeres end in a large duplex loop. *Cell, 97*, 503-514.

Guzman, P. A. & Sanchez, J. G. (1994). Characterization of telomeric regions from Ustilago maydis. *Microbiology, 140 (Pt 3)*, 551-557.

Hardy, C. F., Balderes, D. & Shore, D. (1992). Dissection of a carboxy-terminal region of the yeast regulatory protein RAP1 with effects on both transcriptional activation and silencing. *Mol Cell Biol, 12*, 1209-1217.

Harley, C. B., Futcher, A. B. & Greider, C. W. (1990). Telomeres shorten during ageing of human fibroblasts. *Nature, 345*, 458-460.

Horowitz, H., Thorburn, P. & Haber, J. E. (1984). Rearrangements of highly polymorphic regions near telomeres of Saccharomyces cerevisiae. *Mol Cell Biol, 4*, 2509-2517.

Huang, P., Pryde, F. E., Lester, D., Maddison, R. L., Borts, R. H., Hickson, I. D. & Louis, E. J. (2001). SGS1 is required for telomere elongation in the absence of telomerase. *Curr Biol, 11*, 125-129.

Hug, N. & Lingner, J. (2006). Telomere length homeostasis. *Chromosoma, 115*, 413-425.

Iglesias, N., Redon, S., Pfeiffer, V., Dees, M., Lingner, J. & Luke, B. (2011). Subtelomeric repetitive elements determine TERRA regulation by Rap1/Rif and Rap1/Sir complexes in yeast. *EMBO Rep, 12*, 587-593.

Jaskelioff, M., Muller, F. L., Paik, J. H., Thomas, E., Jiang, S., Adams, A. C., Sahin, E., Kost-Alimova, M., Protopopov, A., Cadinanos, J., Horner, J. W., Maratos-Flier, E. & Depinho, R. A. (2011). Telomerase reactivation reverses tissue degeneration in aged telomerase-deficient mice. *Nature, 469*, 102-106.

Javerzat, J. P., Bhattacherjee, V. & Barreau, C. (1993). Isolation of telomeric DNA from the filamentous fungus Podospora anserina and construction of a self-replicating linear plasmid showing high transformation frequency. *Nucleic Acids Res, 21*, 497-504.

Kamper, J., Kahmann, R., Bolker, M., Ma, L. J., Brefort, T., Saville, B. J., Banuett, F., Kronstad, J. W., Gold, S. E., Muller, O., Perlin, M. H., Wosten, H. A., de Vries, R., Ruiz-Herrera, J., Reynaga-Pena, C. G., Snetselaar, K., McCann, M., Perez-Martin, J., Feldbrugge, M., Basse, C. W., Steinberg, G., Ibeas, J. I., Holloman, W., Guzman, P., Farman, M., Stajich, J. E., Sentandreu, R., Gonzalez-Prieto, J. M., Kennell, J. C., Molina, L., Schirawski, J., Mendoza-Mendoza, A., Greilinger, D., Munch, K., Rossel, N., Scherer, M., Vranes, M., Ladendorf, O., Vincon, V., Fuchs, U., Sandrock, B., Meng, S., Ho, E. C., Cahill, M. J., Boyce, K. J., Klose, J., Klosterman, S. J., Deelstra, H. J., Ortiz-Castellanos, L., Li, W., Sanchez-Alonso, P., Schreier, P. H., Hauser-Hahn, I., Vaupel, M., Koopmann, E., Friedrich, G., Voss, H., Schluter, T., Margolis, J., Platt, D., Swimmer, C., Gnirke, A., Chen, F., Vysotskaia, V., Mannhaupt, G., Guldener, U., Munsterkotter, M., Haase, D., Oesterheld, M., Mewes, H. W., Mauceli, E. W., DeCaprio, D., Wade, C. M., Butler, J., Young, S., Jaffe, D. B., Calvo, S., Nusbaum, C., Galagan, J. & Birren, B. W. (2006). Insights from the genome of the biotrophic fungal plant pathogen Ustilago maydis. *Nature, 444*, 97-101.

Kang, S. S., Kwon, T., Kwon, D. Y. & Do, S. I. (1999). Akt protein kinase enhances human telomerase activity through phosphorylation of telomerase reverse transcriptase subunit. *J Biol Chem, 274*, 13085-13090.

Kelleher, C., Teixeira, M. T., Forstemann, K. & Lingner, J. (2002). Telomerase: biochemical considerations for enzyme and substrate. *Trends Biochem Sci, 27*, 572-579.

Keon, J. P., White, G. A. & Hargreaves, J. A. (1991). Isolation, characterization and sequence of a gene conferring resistance to the systemic fungicide carboxin from the maize smut pathogen, Ustilago maydis. *Curr Genet, 19*, 475-481.

Kim, N. W., Piatyszek, M. A., Prowse, K. R., Harley, C. B., West, M. D., Ho, P. L., Coviello, G. M., Wright, W. E., Weinrich, S. L. & Shay, J. W. (1994). Specific association of human telomerase activity with immortal cells and cancer. *Science, 266*, 2011-2015.

Kimmerly, W. J. & Rine, J. (1987). Replication and segregation of plasmids containing cis-acting regulatory sites of silent mating-type genes in Saccharomyces cerevisiae are controlled by the SIR genes. *Mol Cell Biol, 7*, 4225-4237.

Kimura, A. & Horikoshi, M. (2004). Partition of distinct chromosomal regions: negotiable border and fixed border. *Genes Cells, 9*, 499-508.

Kipling, D. & Cooke, H. J. (1990). Hypervariable ultra-long telomeres in mice. *Nature, 347*, 400-402.

Klobutcher, L. A., Swanton, M. T., Donini, P. & Prescott, D. M. (1981). All gene-sized DNA molecules in four species of hypotrichs have the same terminal sequence and an unusual 3' terminus. *Proc Natl Acad Sci U S A, 78*, 3015-3019.

Kozak, M. L., Chavez, A., Dang, W., Berger, S. L., Ashok, A., Guo, X. & Johnson, F. B. (2010). Inactivation of the Sas2 histone acetyltransferase delays senescence driven by telomere dysfunction. *Embo J, 29*, 158-170.

Kyo, S., Takakura, M., Fujiwara, T. & Inoue, M. (2008). Understanding and exploiting hTERT promoter regulation for diagnosis and treatment of human cancers. *Cancer Sci, 99*, 1528-1538.

Li, H., Zhao, L. L., Funder, J. W. & Liu, J. P. (1997). Protein phosphatase 2A inhibits nuclear telomerase activity in human breast cancer cells. *J Biol Chem, 272*, 16729-16732.

Liu, D., O'Connor, M. S., Qin, J. & Songyang, Z. (2004). Telosome, a mammalian telomere-associated complex formed by multiple telomeric proteins. *J Biol Chem, 279*, 51338-51342.

Longtine, M. S., Enomoto, S., Finstad, S. L. & Berman, J. (1992). Yeast telomere repeat sequence (TRS) improves circular plasmid segregation, and TRS plasmid segregation involves the RAP1 gene product. *Mol Cell Biol, 12*, 1997-2009.

Louis, E. J. & Haber, J. E. (1992). The structure and evolution of subtelomeric Y' repeats in Saccharomyces cerevisiae. *Genetics, 131*, 559-574.

Louis, E. J., Naumova, E. S., Lee, A., Naumov, G. & Haber, J. E. (1994). The chromosome end in yeast: its mosaic nature and influence on recombinational dynamics. *Genetics, 136*, 789-802.

Luke, B., Panza, A., Redon, S., Iglesias, N., Li, Z. & Lingner, J. (2008). The Rat1p 5' to 3' exonuclease degrades telomeric repeat-containing RNA and promotes telomere elongation in Saccharomyces cerevisiae. *Mol Cell, 32*, 465-477.

Lundblad, V. & Blackburn, E. H. (1993). An alternative pathway for yeast telomere maintenance rescues est1- senescence. *Cell, 73*, 347-360.

Lundblad, V. & Szostak, J. W. (1989). A mutant with a defect in telomere elongation leads to senescence in yeast. *Cell, 57*, 633-643.

Lydall, D. (2009). Taming the tiger by the tail: modulation of DNA damage responses by telomeres. *Embo J, 28*, 2174-2187.

Maida, Y., Yasukawa, M., Furuuchi, M., Lassmann, T., Possemato, R., Okamoto, N., Kasim, V., Hayashizaki, Y., Hahn, W. C. & Masutomi, K. (2009). An RNA-dependent RNA polymerase formed by TERT and the RMRP RNA. *Nature, 461*, 230-235.

Mandell, J. G., Goodrich, K. J., Bahler, J. & Cech, T. R. (2005). Expression of a RecQ helicase homolog affects progression through crisis in fission yeast lacking telomerase. *J Biol Chem, 280*, 5249-5257.

Mao, N., Kojic, M. & Holloman, W. K. (2009). Role of Blm and collaborating factors in recombination and survival following replication stress in Ustilago maydis. *DNA Repair (Amst), 8*, 752-759.

Masutomi, K., Yu, E. Y., Khurts, S., Ben-Porath, I., Currier, J. L., Metz, G. B., Brooks, M. W., Kaneko, S., Murakami, S., DeCaprio, J. A., Weinberg, R. A., Stewart, S. A. & Hahn, W. C. (2003). Telomerase maintains telomere structure in normal human cells. *Cell, 114*, 241-253.

Moxon, E. R., Lenski, R. E. & Rainey, P. B. (1998). Adaptive evolution of highly mutable loci in pathogenic bacteria. *Perspect Biol Med, 42*, 154-155.

Munoz-Jordan, J. L., Cross, G. A., de Lange, T. & Griffith, J. D. (2001). t-loops at trypanosome telomeres. *Embo J, 20*, 579-588.

Oganesian, L. & Karlseder, J. (2010). Mammalian 5' C-rich telomeric overhangs are a mark of recombination-dependent telomere maintenance. *Mol Cell, 42*, 224-236.

Palm, W. & de Lange, T. (2008). How shelterin protects mammalian telomeres. *Annu Rev Genet, 42*, 301-334.

Park, J. I., Venteicher, A. S., Hong, J. Y., Choi, J., Jun, S., Shkreli, M., Chang, W., Meng, Z., Cheung, P., Ji, H., McLaughlin, M., Veenstra, T. D., Nusse, R., McCrea, P. D. & Artandi, S. E. (2009). Telomerase modulates Wnt signalling by association with target gene chromatin. *Nature, 460*, 66-72.

Perez, G., Pangilinan, J., Pisabarro, A. G. & Ramirez, L. (2009). Telomere organization in the ligninolytic basidiomycete Pleurotus ostreatus. *Appl Environ Microbiol, 75*, 1427-1436.

Power, P., Jeffery, D., Rehman, M. A., Chatterji, A. & Yankulov, K. (2011). Sub-telomeric core X and Y' elements in S. cerevisiae suppress extreme variations in gene silencing. *PLoS One, 6*, e17523.

Price, C. M., Boltz, K. A., Chaiken, M. F., Stewart, J. A., Beilstein, M. A. & Shippen, D. E. (2010). Evolution of CST function in telomere maintenance. *Cell Cycle, 9*, 3157-3165.

Pryde, F. E. & Louis, E. J. (1999). Limitations of silencing at native yeast telomeres. *Embo J, 18*, 2538-2550.

Raices, M., Verdun, R. E., Compton, S. A., Haggblom, C. I., Griffith, J. D., Dillin, A. & Karlseder, J. (2008). C. elegans telomeres contain G-strand and C-strand overhangs that are bound by distinct proteins. *Cell, 132*, 745-757.

Rehmeyer, C. J., Li, W., Kusaba, M. & Farman, M. L. (2009). The telomere-linked helicase (TLH) gene family in Magnaporthe oryzae: revised gene structure reveals a novel TLH-specific protein motif. *Current genetics, 55*, 253-262.

Sahin, E., Colla, S., Liesa, M., Moslehi, J., Muller, F. L., Guo, M., Cooper, M., Kotton, D., Fabian, A. J., Walkey, C., Maser, R. S., Tonon, G., Foerster, F., Xiong, R., Wang, Y. A., Shukla, S. A., Jaskelioff, M., Martin, E. S., Heffernan, T. P., Protopopov, A., Ivanova, E., Mahoney, J. E., Kost-Alimova, M., Perry, S. R., Bronson, R., Liao, R., Mulligan, R., Shirihai, O. S., Chin, L. & DePinho, R. A. (2011). Telomere dysfunction induces metabolic and mitochondrial compromise. *Nature, 470*, 359-365.

Sanchez-Alonso, P. & Guzman, P. (1998). Organization of chromosome ends in Ustilago maydis. RecQ-like helicase motifs at telomeric regions. *Genetics, 148*, 1043-1054.

Sanchez-Alonso, P. & Guzman, P. (2008). Predicted elements of telomere organization and function in Ustilago maydis. *Fungal Genet Biol, 45 Suppl 1*, S54-62.

Sanchez-Alonso, P., Valverde, M. E., Paredes-Lopez, O. & Guzman, P. (1996). Detection of genetic variation in Ustilago maydis strains by probes derived from telomeric sequences. *Microbiology, 142 (Pt 10)*, 2931-2936.

Schoeftner, S. & Blasco, M. A. (2008). Developmentally regulated transcription of mammalian telomeres by DNA-dependent RNA polymerase II. *Nat Cell Biol, 10*, 228-236.

Schoeftner, S. & Blasco, M. A. (2010). Chromatin regulation and non-coding RNAs at mammalian telomeres. *Semin Cell Dev Biol, 21*, 186-193.

Smith, J. J., Miller, L. R., Kreisberg, R., Vazquez, L., Wan, Y. & Aitchison, J. D. (2011). Environment-responsive transcription factors bind subtelomeric elements and regulate gene silencing. *Mol Syst Biol, 7*, 455.

Takahashi, K., Imano, R., Kibe, T., Seimiya, H., Muramatsu, Y., Kawabata, N., Tanaka, G., Matsumoto, Y., Hiromoto, T., Koizumi, Y., Nakazawa, N., Yanagida, M., Yukawa, M., Tsuchiya, E. & Ueno, M. (2011). Fission yeast Pot1 and RecQ helicase are required for efficient chromosome segregation. *Molecular and cellular biology, 31*, 495-506.

Teixeira, M. T., Arneric, M., Sperisen, P. & Lingner, J. (2004). Telomere length homeostasis is achieved via a switch between telomerase- extendible and -nonextendible states. *Cell, 117*, 323-335.

Teng, S. C. & Zakian, V. A. (1999). Telomere-telomere recombination is an efficient bypass pathway for telomere maintenance in Saccharomyces cerevisiae. *Mol Cell Biol, 19*, 8083-8093.

Teunissen, A. W. & Steensma, H. Y. (1995). Review: the dominant flocculation genes of Saccharomyces cerevisiae constitute a new subtelomeric gene family. *Yeast, 11*, 1001-1013.

Tham, W. H. & Zakian, V. A. (2002). Transcriptional silencing at Saccharomyces telomeres: implications for other organisms. *Oncogene, 21*, 512-521.

Ulaner, G. A., Hu, J. F., Vu, T. H., Oruganti, H., Giudice, L. C. & Hoffman, A. R. (2000). Regulation of telomerase by alternate splicing of human telomerase reverse transcriptase (hTERT) in normal and neoplastic ovary, endometrium and myometrium. *Int J Cancer, 85*, 330-335.

van Overveld, P. G., Lemmers, R. J., Deidda, G., Sandkuijl, L., Padberg, G. W., Frants, R. R. & van der Maarel, S. M. (2000). Interchromosomal repeat array interactions between chromosomes 4 and 10: a model for subtelomeric plasticity. *Hum Mol Genet, 9*, 2879-2884.

Vindigni, A. & Hickson, I. D. (2009). RecQ helicases: multiple structures for multiple functions? *Hfsp J, 3*, 153-164.

Vodenicharov, M. D. & Wellinger, R. J. (2006). DNA degradation at unprotected telomeres in yeast is regulated by the CDK1 (Cdc28/Clb) cell-cycle kinase. *Mol Cell, 24*, 127-137.

Walbot, V. & Skibbe, D. S. (2010). Maize host requirements for Ustilago maydis tumor induction. *Sex Plant Reprod, 23*, 1-13.

Wang, R. C., Smogorzewska, A. & de Lange, T. (2004). Homologous recombination generates T-loop-sized deletions at human telomeres. *Cell, 119*, 355-368.

Williams, R. S., Williams, J. S. & Tainer, J. A. (2007). Mre11-Rad50-Nbs1 is a keystone complex connecting DNA repair machinery, double-strand break signaling, and the chromatin template. *Biochem Cell Biol, 85*, 509-520.

Wong, J. M. & Collins, K. (2006). Telomerase RNA level limits telomere maintenance in X-linked dyskeratosis congenita. *Genes Dev, 20*, 2848-2858.

Wright, J. H., Gottschling, D. E. & Zakian, V. A. (1992). Saccharomyces telomeres assume a non-nucleosomal chromatin structure. *Genes Dev, 6*, 197-210.

Yamada, M., Hayatsu, N., Matsuura, A. & Ishikawa, F. (1998). Y'-Help1, a DNA helicase encoded by the yeast subtelomeric Y' element, is induced in survivors defective for telomerase. *J Biol Chem, 273*, 33360-33366.

Yang, C., Przyborski, S., Cooke, M. J., Zhang, X., Stewart, R., Anyfantis, G., Atkinson, S. P., Saretzki, G., Armstrong, L. & Lako, M. (2008). A key role for telomerase reverse transcriptase unit in modulating human embryonic stem cell proliferation, cell cycle dynamics, and in vitro differentiation. *Stem Cells, 26*, 850-863.

Yehezkel, S., Rebibo-Sabbah, A., Segev, Y., Tzukerman, M., Shaked, R., Huber, I., Gepstein, L., Skorecki, K. & Selig, S. (2011). Reprogramming of telomeric regions during the generation of human induced pluripotent stem cells and subsequent differentiation into fibroblast-like derivatives. *Epigenetics, 6*, 63-75.

Yu, C. E., Oshima, J., Fu, Y. H., Wijsman, E. M., Hisama, F., Alisch, R., Matthews, S., Nakura, J., Miki, T., Ouais, S., Martin, G. M., Mulligan, J. & Schellenberg, G. D. (1996). Positional cloning of the Werner's syndrome gene. *Science, 272*, 258-262.

Zvereva, M. I., Shcherbakova, D. M. & Dontsova, O. A. (2010). Telomerase: structure, functions, and activity regulation. *Biochemistry (Mosc), 75*, 1563-1583.

In: Advances in Genetics Research. Volume 12
Editor: Kevin V. Urbano
ISBN: 978-1-63321-351-7

Chapter 7

ROLES OF miRNAs IN CANCERS

Shigeru Takasaki*
Toyo University, Ora-gun Gunma, Japan

ABSTRACT

MicroRNAs (miRNAs) are small (~24 nucleotides) noncoding RNA molecules thought to play an important role in regulating gene expression. Although knowledge of the biological functions of most miRNAs is still limited, miRNAs are thought to regulate the gene expression in various diseases. In the article, the roles of miRNAs in human cancers are examined from the viewpoint of up- and down-regulation. Oncogenic miRNAs are involved in the overexpression/upregulation of cancers, whereas suppressive miRNAs are involved in the underexpression/downregulation of cancers. The oncogenic and suppressive miRNA sequences were first collected from miRBase based on the relations between miRNAs and various cancers in the literature. Statistical analysis of the positional nucleotide occurrence features of miRNAs revealed differences between the positional nucleotide occurrences of oncogenic and suppressive miRNAs. A miRNA gene-silencing score was then defined on the basis of the higher and lower levels of the statistical significances of positional nucleotides. Since the miRNA scores were closely related to miRNA frequencies, a method using the scores and nucleotide frequencies to distinguish whether a new miRNA is oncogenic or suppressive is proposed.

Keywords: miRNA, noncoding RNA, gene silencing, cancer, oncogenic, suppressive, significance test

INTRODUCTION

MicroRNAs (miRNAs) are small (~24 nucleotides) noncoding RNA molecules that regulate gene expression post-transcriptionally by base-pairing to mRNAs [1–4]. Many miRNAs have recently been identified in various multicellular organisms and are

* Correspondence should be addressed to s_takasaki@toyo.jp. Phone: +81-276-82-9024. Fax: +81-276-82-9033.

evolutionally conserved. Knowledge of the biological functions of most miRNAs is still limited, but these molecules are thought to regulate the gene expression at various stages in diseases and development.

miRNA Biogenesis Pathway

Animal microRNAs are typically transcribed as primary transcripts (pri-miRNAs) of varying length that in the nucleus are processed by Drosha into stem-loop precursors consisting of ~70 nucleotides. In the cytoplasm the precursor miRNA (pre-miRNA) is processed by the RNase III type enzyme Dicer to give a mature ~24-nucleotide product. MicroRNAs regulate the expression levels of other genes by several mechanisms, generally reducing protein levels through the cleavage of mRNAs or the repression of their translation.

The pathway of miRNA biogenesis is shown in Figure 1, and the numbered steps there are described as follows:

1. Pri-miRNA is transcribed by RNA polymerase II or III. The length of the pri-miRNA ranges from several hundred nucleotides to thousands of nucleotides.
2. The pri-miRNA is cleaved into pre-miRNA hairpins by the Drosha/DGC8 complex in the nucleus.
3. The pre-miRNA is transferred from the nucleus to the cytoplasm by Exportin-5.
4. The pre-miRNA in the cytoplasm is cleaved by the RNase III enzyme Dicer to give a mature ~24-nucleotide product.
5. A guide miRNA, the functional strand of the mature miRNA, is formed by the RNA-induced silencing complex (RISC).
6. The guide miRNA is incorporated by the RISC into the target mRNA, which it silences through deadenylation, cleavage, or transcriptional repression.

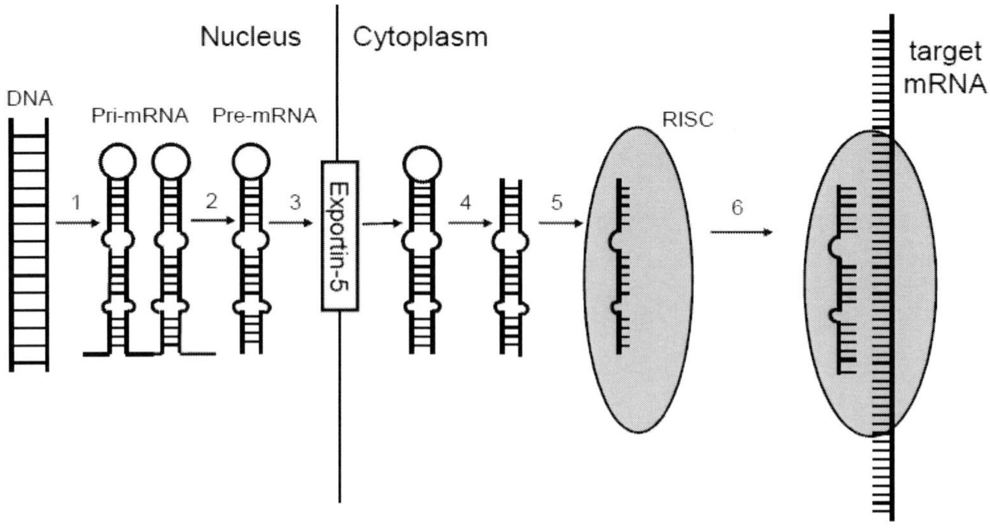

Figure 1. miRNA biogenesis pathway.

ROLES OF miRNAS IN CANCERS

The Relations between miRNAs and Human Cancers

It has been observed that miRNAs contribute to cancer development and progression. Croce and his colleagues analyzed 540 samples of lung, breast, stomach, prostate, colon, and pancreatic tumors and found that miRNAs are differentially expressed in normal tissues and cancers [5]. Other studies have also reported that miRNAs play some roles as oncogenes or tumor suppressor genes. They can therefore be used as biomarkers in human cancers [6–41]. In Table 1 miRNAs that are upregulated in various cancers are listed as oncogenic miRNAs and miRNAs that are downregulated in those cancers are listed as suppressive miRNAs [6–41]. It is clear from Table 1 that there are differences in occurrences of the oncogenic and suppressive miRNAs for the individual cancers. The related individual miRNA frequencies are listed in Tables 2(A) and 2(B). More miRNAs are oncogenic than are suppressive. hsa-miR-21, for example, is oncogenic for ten cancers. Other miRNAs related oncogenic for more than four cancers are hsa-miR-17, -191, -221, -155, -214, -223, -106a, -199a and -92a. In contrast, only hsa-miR-145, let-7, and hsa-miR-143 are suppressive for more than five kinds of cancers.

Table 1. Relations between human cancers and miRNAs

cancer entity	oncogenic miRNAs	suppressive miRNAs
brain, GBM	hsa-miR-21, 221	hsa-miR-128, 181
breast	hsa-miR-9, 10b, 17, 21, 29b, 34, 146, 155, 181b, 181a	hsa-let-7, hsa-miR-15a, 16, 125a, 125b, 127, 145, 204
lung	hsa-miR-17, 21, 24-2, 106a, 128, 146, 150, 155, 191, 192, 197, 199a, 205, 212, 210, 214	hsa-let-7, hsa-miR-9, 26a, 27b, 29b, 32, 34, 30a, 95, 101, 124, 125a, 126, 140, 143, 145, 181c, 192, 198, 199b, 216, 218, 219, 224
esophagus	hsa-miR-21, 93	hsa-miR-203, 205
stomach	hsa-miR-21, 24, 25, 92a, 107, 191, 214, 221, 223	hsa-let-7
colorectal	hsa-miR-17, 20a, 21, 24, 29b, 30c, 31, 32, 96, 106a, 107, 128b, 135b, 155, 183, 191, 221, 223	hsa-let-7, has-miR-34, 127, 133b, 143, 145
hepatic cell	hsa-miR-15b, 18a, 21, 106b, 221, 222, 224	hsa-let-7, has-miR-101, 122, 125a, 195, 199a, 200a
pancreas	hsa-miR-17, 20a, 21, 24, 25, 29b, 30c, 32, 92a, 100, 106a, 107, 125b, 128, 146, 155, 181a, 181b, 191, 196a, 196b, 199a, 214, 221, 223, 301a, 376a	hsa-miR-139, 142, 345, 375
prostate gland	hsa-miR-17, 20a, 21, 25, 30c, 32,92a, 106a, 146a, 181b, 191, 199a, 214, 223	hsa-miR-15a, 16, 143, 145, 218
cervical	hsa-miR-21, 199a	hsa-miR-143, 145
B-CLL	hsa-miR-17, 92a-1	hsa-miR-15a, 16, 143, 145, 192, 181a

GBM: glioblastoma maltiforme, B-CLL: B cell chromic lymphocytic leukemia.

Table 2(A). miRNA frequencies of the oncogenic miRNAs for various cancers

miRNA	frequency	cancer type										
		1	2	3	4	5	6	7	8	9	10	11
hsa-miR-21	10	O	O	O	O	O	O	O	O	O	O	
-17	6		O	O			O		O	O		O
-191	5			O		O	O		O	O		
-221	5	O				O	O	O	O			
-155	4		O	O			O		O			
-214	4			O		O			O	O		
-223	4					O	O		O	O		
-106a	4			O			O		O	O		
-199a	4			O					O	O	O	
-92a	4					O			O	O		O
-24	3					O	O		O			
-25	3					O			O	O		
-32	3						O		O	O		
-107	3					O	O		O			
-146	3		O	O					O			
-181b	3		O						O	O		
-20a	3						O		O	O		
-29b	3		O				O		O			
-30c	3						O		O	O		
-128	2			O					O			
-181a	2		O						O			

1: brain, GBM, 2:breast, 3:lung, 4: esophagus, 5: stomach,6: colorectal, 7: hepatic cell, 8: pancreas, 9: prostate gland, 10: cervical, 11: B-CLL

Table 2(B). miRNA frequencies of the suppressive miRNAs for various cancers

miRNA	frequency	cancer type										
		1	2	3	4	5	6	7	8	9	10	11
hsa-miR-145	6		O	O			O			O	O	O
let-7	5		O	O		O	O	O				
hsa-miR-143	5			O			O			O	O	O
-16	3		O							O		O
-125a	3		O	O				O				
-15a	3		O							O		O
-32	2			O						O		
-34	2			O			O					
-101	2			O				O				
-127	2		O				O					
-192	2			O								O
-218	2			O						O		

1: brain, GBM, 2:breast, 3:lung, 4: esophagus, 5: stomach,6: colorectal, 7: hepatic cell, 8: pancreas, 9: prostate gland, 10: cervical, 11: B-CLL

On the other hand, although RNA interference (RNAi) has been widely used for studying gene functions, the gene-silencing effectiveness in mammalian cells depends very much on target sequence positions (sites) selected from the target gene. That is, different short interfering RNAs (siRNAs) from various positions induce different levels of gene-silencing. The gene-silencing of miRNAs may also be considered to depend on the selected target sequence positions. Since the target mRNA sequence is a certain combination of nucleotides, some nucleotide sequences of miRNAs can be considered to play roles in the gene-silencing of the target mRNA.

Positional Nucleotide Features of Oncogenic and Suppressive miRNAs

To get positional nucleotide features, the nucleotide sequences for the oncogenic and suppressive miRNAs listed in Table 1 were first collected from miRBase [42, 43]. The obtained miRNA sequences are listed in Tables 3(A) and 3(B). Then the frequencies of the four nucleotides (A, G, C, U) at positions from 1 to 22 were determined for the pooled sequences.

These frequencies are listed in Tables 4(A) and 4(B). As shown in Figs. 2(A) and 2(B), there are differences in nucleotide distribution ratios at individual positions for oncogenic and suppressive miRNAs. (As there are a few nucleotide occurrences at position 23 listed in Tables 3(A) and 3(B), the frequencies of the nucleotides at positions from 1 to 22 are used in Tables 4(A) and 4(B) and Figs. 2(A) and 2(B).)

Table 3(A). Nucleotide sequences of oncogenic miRNAs for various cancers

cancer entity	has-miR-	1	2	3	4	5	6	7	8	9	10	11	12	13	14	15	16	17	18	19	20	21	22	23
brain, GBM	21-5p	U	A	G	C	U	U	A	U	C	A	G	A	C	U	G	A	U	G	U	U	G	A	
	-3p	C	A	A	C	A	C	C	A	G	U	C	G	A	U	G	G	G	U	U	G	U		
	221-5p	A	C	C	U	G	G	C	A	U	A	C	A	A	U	G	U	A	G	A	U	U	U	
breast	9-5p	U	C	U	U	U	G	G	U	U	A	U	C	U	A	G	C	U	G	U	A	U	G	A
	9-3p	A	U	A	A	A	G	C	U	A	G	A	U	A	A	C	C	G	A	A	A	G	U	
	10b-5p	U	A	C	C	C	U	G	U	A	G	A	A	C	C	G	A	A	U	U	U	G	U	G
	10p-3p	A	C	A	G	A	U	U	C	G	A	U	U	C	U	A	G	G	G	G	A	A	U	
	17-5p	C	A	A	A	G	U	G	C	U	U	A	C	A	G	U	G	C	A	G	G	U	A	G
	21-5p	U	A	G	C	U	U	A	U	C	A	G	A	C	U	G	A	U	G	U	U	G	A	
	21-3p	C	A	A	C	A	C	C	A	G	U	C	G	A	U	G	G	G	C	U	G	U		
	29b-2-5p	C	U	G	G	U	U	U	C	A	C	A	U	G	G	U	G	G	C	U	U	A	G	
	34a-5p	U	G	G	C	A	G	U	G	U	C	U	U	A	G	C	U	G	G	U	U	G	U	
	34a-3p	C	A	A	U	C	A	G	C	A	A	G	U	A	U	A	C	U	G	C	C	C	U	
	34b-5p	U	A	G	G	C	A	G	U	G	U	C	A	U	U	A	G	C	U	G	A	U	U	G
	34b-3p	C	A	A	U	C	A	C	U	A	A	C	U	C	C	A	C	U	G	C	C	A	U	
	34c-5p	A	G	G	C	A	G	U	G	U	A	G	U	U	A	G	C	U	G	A	U	U	G	C
	34c-3p	A	A	U	C	A	C	U	A	A	C	C	A	C	A	C	G	G	C	C	A	G	G	
	146a-5p	U	G	A	G	A	A	C	U	G	A	A	U	U	C	C	A	U	G	G	G	U	U	
	146b-5p	U	G	A	G	A	A	C	U	G	A	A	U	U	C	C	A	U	A	G	G	C	U	
	155-5p	U	U	A	A	U	G	C	U	A	A	U	C	G	U	G	A	U	A	G	G	G	G	U
	181b-5p	A	A	C	A	U	U	C	A	U	U	G	C	U	G	U	C	G	G	U	G	G	G	U
	181a-5p	A	A	C	A	U	U	C	A	A	C	G	C	U	G	U	C	G	G	U	G	A	G	U
	181a-3p	A	C	C	A	U	C	G	A	C	C	G	U	U	G	A	U	U	G	U	A	C	C	

Table 3(A). (Continued)

cancer entity	has-miR-	1	2	3	4	5	6	7	8	9	10	11	12	13	14	15	16	17	18	19	20	21	22	23
lung	17-5p	C	A	A	A	G	U	G	C	U	U	A	C	A	G	U	G	C	A	G	G	U	A	G
	21-5p	U	A	G	C	U	U	A	U	C	A	G	A	C	U	G	A	U	G	U	U	G	A	
	24-2-5p	U	G	C	C	U	A	C	U	G	A	G	C	U	G	A	A	A	C	A	C	A	G	
	106a-5p	A	A	A	A	G	U	G	C	U	U	A	C	A	G	U	G	C	A	G	G	U	A	G
	128-3p	U	C	A	C	A	G	U	G	A	A	C	C	G	G	U	C	U	C	U	U	U		
	146a-5p	U	G	A	G	A	A	C	U	G	A	A	U	U	C	C	A	U	G	G	G	U	U	
	146b-5p	U	G	A	G	A	A	C	U	G	A	A	U	U	C	C	A	U	A	G	G	C	U	
	150-5p	U	C	U	C	C	C	A	A	C	C	C	U	U	G	U	A	C	C	A	G	U	G	
	155-5p	U	U	A	A	U	G	C	U	A	A	U	C	G	U	G	A	U	A	G	G	G	G	U
	191-5p	C	A	A	C	G	G	A	A	U	C	C	C	A	A	A	A	G	C	A	G	C	U	G
	192-5p	C	U	G	A	C	C	U	A	U	G	A	A	U	U	G	A	C	A	G	C	C		
	197-3p	U	U	C	A	C	C	A	C	C	U	U	C	U	C	C	A	C	C	C	A	G	C	
	197-5p	C	G	G	G	U	A	G	A	G	A	G	G	G	C	A	G	U	G	G	G	A	G	G
	199a-5p	C	C	C	A	G	U	G	U	U	C	A	G	A	C	U	A	C	C	U	G	U	U	C
	205-5p	U	C	C	U	U	C	A	U	U	C	C	A	C	C	G	G	A	G	U	C	U	G	
	212-3p	U	A	A	C	A	G	U	C	U	C	C	A	G	U	C	A	C	G	G	C	C		
	212-5p	A	C	C	U	U	G	G	C	U	C	U	A	G	A	C	U	G	C	U	U	A	C	U
	210-3p	C	U	G	U	G	C	G	U	G	U	G	A	C	A	G	C	G	G	C	U	G	A	
	214-3p	A	C	A	G	C	A	G	G	C	A	C	A	G	A	C	A	G	G	C	A	G	U	
esophagus	21-5p	U	A	G	C	U	U	A	U	C	A	G	A	C	U	G	A	U	G	U	U	G	A	
	93-5p	C	A	A	A	G	U	G	C	U	G	U	U	C	G	U	G	C	A	G	G	U	A	G
stomach	21-5p	U	A	G	C	U	U	A	U	C	A	G	A	C	U	G	A	U	G	U	U	G	A	
	24-3p	U	G	G	C	U	C	A	G	U	U	C	A	G	C	A	G	G	A	A	C	A	G	
	25-5p	A	G	G	C	G	G	A	G	A	C	U	U	G	G	G	C	A	A	U	U	G		
	25-3p	C	A	U	U	G	C	A	C	U	U	G	U	C	U	C	G	G	U	C	U	G	A	
	92a-3p	U	A	U	U	G	C	A	C	U	U	G	U	C	C	C	G	G	C	C	U	G	U	
	107	A	G	C	A	G	C	A	U	U	G	U	A	C	A	G	G	G	C	U	A	U	C	A
	191-5p	C	A	A	C	G	G	A	A	U	C	C	C	A	A	A	A	G	C	A	G	C	U	G
	214-3p	A	C	A	G	C	A	G	G	C	A	C	A	G	A	C	A	G	G	C	A	G	U	
	221-5p	A	C	C	U	G	G	C	A	U	A	C	A	A	U	G	U	A	G	A	U	U	U	
	221-3p	A	G	C	U	A	C	A	U	U	G	U	C	U	G	C	U	G	G	G	U	U	U	C
	223-3p	U	G	U	C	A	G	U	U	U	G	U	C	A	A	A	U	A	C	C	C	C	A	
colorectal	17-5p	C	A	A	A	G	U	G	C	U	U	A	C	A	G	U	G	C	A	G	G	U	A	G
	20a-5p	U	A	A	A	G	U	G	C	U	U	A	U	A	G	U	G	C	A	G	G	U	A	G
	21-5p	U	A	G	C	U	U	A	U	C	A	G	A	C	U	G	A	U	G	U	U	G	A	
	24-3p	U	G	G	C	U	C	A	G	U	U	C	A	G	C	A	G	G	A	A	C	A	G	
	29b-3p	U	A	G	C	A	C	C	A	U	U	U	G	A	A	A	U	C	A	G	U	G	U	U
	30c-5p	U	G	U	A	A	A	C	A	U	C	C	U	A	C	A	C	U	C	U	C	A	G	C
	31-5p	A	G	G	C	A	A	G	A	U	G	C	U	G	G	C	A	U	A	G	C	U		
	32-5p	U	A	U	U	G	C	A	C	A	U	U	A	C	U	A	A	G	U	U	G	C	A	
	96-5p	U	U	U	G	G	C	A	C	U	A	G	C	A	C	A	U	U	U	U	U	G	C	U
	106a-5p	A	A	A	A	G	U	G	C	U	U	A	C	A	G	U	G	C	A	G	G	U	A	G
	107	A	G	C	A	G	C	A	U	U	G	U	A	C	A	G	G	G	C	U	A	U	C	A
	128-3p	U	C	A	C	A	G	U	G	A	A	C	C	G	G	U	C	U	C	U	U	U		
	135b-5p	U	A	U	G	G	C	U	U	U	U	C	A	U	U	C	C	U	A	U	G	U	G	A
	155-5p	U	U	A	A	U	G	C	U	A	A	U	C	G	U	G	A	U	A	G	G	G	G	U
	183-5p	U	A	U	G	G	C	A	C	U	G	G	U	A	G	A	A	U	U	C	A	C	U	
	191-5p	C	A	A	C	G	G	A	A	U	C	C	C	A	A	A	A	G	C	A	G	C	U	G
	221-5p	A	C	C	U	G	G	C	A	U	A	C	A	A	U	G	U	A	G	A	U	U	U	
	221-3p	A	G	C	U	A	C	A	U	U	G	U	C	U	G	C	U	G	G	G	U	U	U	C
	223-3p	U	G	U	C	A	G	U	U	U	G	U	C	A	A	A	U	A	C	C	C	C	A	

cancer entity	has-miR-	1	2	3	4	5	6	7	8	9	10	11	12	13	14	15	16	17	18	19	20	21	22	23
hepatic cell	15b-5p	U	A	G	C	A	G	C	A	C	A	U	C	A	U	G	G	U	U	U	A	C	A	
	18a-5p	U	A	A	G	G	U	G	C	A	U	C	U	A	G	U	G	C	A	G	A	U	A	G
	21-5p	U	A	G	C	U	U	A	U	C	A	G	A	C	U	G	A	U	G	U	U	G	A	
	106b-5p	U	A	A	A	G	U	G	C	U	G	A	C	A	G	U	G	C	A	G	A	U		
	106b-3p	C	C	G	C	A	C	U	G	U	G	G	G	U	A	C	U	U	G	C	U	G	C	
	221-5p	A	C	C	U	G	G	C	A	U	A	C	A	A	U	G	U	A	G	A	U	U	U	
	221-3p	A	G	C	U	A	C	A	U	U	G	U	C	U	G	C	U	G	G	G	U	U	U	C
	222-3p	A	G	C	U	A	C	A	U	C	U	G	G	C	U	A	C	U	G	G	G	U		
	224-5p	C	A	A	G	U	C	A	C	U	A	G	U	G	G	U	U	C	C	G	U	U		
pancreas	17-5p	C	A	A	A	G	U	G	C	U	U	A	C	A	G	U	G	C	A	G	G	U	A	G
	20a-5p	U	A	A	A	G	U	G	C	U	U	A	U	A	G	U	G	C	A	G	G	U	A	G
	21-5p	U	A	G	C	U	U	A	U	C	A	G	A	C	U	G	A	U	G	U	U	G	A	
	24-3p	U	G	G	C	U	C	A	G	U	U	C	A	G	C	A	G	G	A	A	C	A	G	
	25-5p	A	G	G	C	G	G	A	G	A	C	U	U	G	G	G	C	A	A	U	U	G		
	29b-3p	U	A	G	C	A	C	C	A	U	U	U	G	A	A	A	U	C	A	G	U	G	U	U
	30c-5p	U	G	U	A	A	A	C	A	U	C	C	U	A	C	A	C	U	C	U	C	A	G	C
	32-5p	U	A	U	U	G	C	A	C	A	U	U	A	C	U	A	A	G	U	U	G	C	A	
	92a-3p	U	A	U	U	G	C	A	C	U	U	G	U	C	C	C	G	G	C	C	U	G	U	
	100-5p	A	A	C	C	C	G	U	A	G	A	U	C	C	G	A	A	C	U	U	G	U	G	
	106a-5p	A	A	A	A	G	U	G	C	U	U	A	C	A	G	U	G	C	A	G	G	U	A	G
	107	A	G	C	A	G	C	A	U	U	G	U	A	C	A	G	G	G	C	U	A	U	C	A
	125b-5p	U	C	C	C	U	G	A	G	A	C	C	C	U	A	A	C	U	U	G	U	G	A	
	128-3p	U	C	A	C	A	G	U	G	A	A	C	C	G	G	U	C	U	C	U	U	U		
	146a-5p	U	G	A	G	A	A	C	U	G	A	A	U	U	C	C	A	U	G	G	G	U	U	
	155-5p	U	U	A	A	U	G	C	U	A	A	U	C	G	U	G	A	U	A	G	G	G	G	U
	181a-5p	A	A	C	A	U	U	C	A	A	C	G	C	U	G	U	C	G	G	U	G	A	G	U
	181a-3p	A	C	C	A	U	C	G	A	C	C	G	U	U	G	A	U	U	G	U	A	C	C	
	181b-5p	A	A	C	A	U	U	C	A	U	U	G	C	U	G	U	C	G	G	U	G	G	G	U
	191-5p	C	A	A	C	G	G	A	A	U	C	C	C	A	A	A	A	G	C	A	G	C	U	G
	196a-5p	U	A	G	G	U	A	G	U	U	U	C	A	U	G	U	U	G	U	U	G	G	G	
	199a-5p	C	C	C	A	G	U	G	U	U	C	A	G	A	C	U	A	C	C	U	G	U	U	C
	199a-3p	A	C	A	G	U	A	G	U	C	U	G	C	A	C	A	U	U	G	G	U	U	A	
	214-3p	A	C	A	G	C	A	G	G	C	A	C	A	G	A	C	A	G	G	C	A	G	U	
	221-5p	A	C	C	U	G	G	C	A	U	A	C	A	A	U	G	U	A	G	A	U	U	U	
	221-3p	A	G	C	U	A	C	A	U	U	G	U	C	U	G	C	U	G	G	G	U	U	U	C
	223-3p	U	G	U	C	A	G	U	U	U	G	U	C	A	A	A	U	A	C	C	C	C	A	
	301a-5p	G	C	U	C	U	G	A	C	U	U	U	A	U	U	G	C	A	C	U	A	C	U	
	301a-3p	C	A	G	U	G	C	A	A	U	A	G	U	A	U	U	G	U	C	A	A	A	G	C
	376a-3p	A	U	C	A	U	A	G	A	G	G	A	A	A	A	U	C	C	A	C	G	U		
prostate gland	17-5p	C	A	A	A	G	U	G	C	U	U	A	C	A	G	U	G	C	A	G	G	U	A	G
	20a-5p	U	A	A	A	G	U	G	C	U	U	A	U	A	G	U	G	C	A	G	G	U	A	G
	21-5p	U	A	G	C	U	U	A	U	C	A	G	A	C	U	G	A	U	G	U	U	G	A	
	25-5p	A	G	G	C	G	G	A	G	A	C	U	U	G	G	G	C	A	A	U	U	G		
	30c-5p	U	G	U	A	A	A	C	A	U	C	C	U	A	C	A	C	U	C	U	C	A	G	C
	32-5p	U	A	U	U	G	C	A	C	A	U	U	A	C	U	A	A	G	U	U	G	C	A	
	92a-3p	U	A	U	U	G	C	A	C	U	U	G	U	C	C	C	G	G	C	C	U	G	U	
	106a-5p	A	A	A	A	G	U	G	C	U	U	A	C	A	G	U	G	C	A	G	G	U	A	G
	146a-5p	U	G	A	G	A	A	C	U	G	A	A	U	U	C	C	A	U	G	G	G	U	U	
	181b-5p	A	A	C	A	U	U	C	A	U	U	G	C	U	G	U	C	G	G	U	G	G	G	U
	191-5p	C	A	A	C	G	G	A	A	U	C	C	C	A	A	A	A	G	C	A	G	C	U	G
	199a-5p	C	C	C	A	G	U	G	U	U	C	A	G	A	C	U	A	C	C	U	G	U	U	C
	221-3p	A	G	C	U	A	C	A	U	U	G	U	C	U	G	C	U	G	G	G	U	U	U	C
	223-3p	U	G	U	C	A	G	U	U	U	G	U	C	A	A	A	U	A	C	C	C	C	A	
cervical	21-5p	U	A	G	C	U	U	A	U	C	A	G	A	C	U	G	A	U	G	U	U	G	A	
	199a-5p	C	C	C	A	G	U	G	U	U	C	A	G	A	C	U	A	C	C	U	G	U	U	C
B-CELL	17-5p	C	A	A	A	G	U	G	C	U	U	A	C	A	G	U	G	C	A	G	G	U	A	G
	92a-1-5p	A	G	G	U	U	G	G	G	A	U	C	G	G	U	U	G	C	A	A	U	G	C	U
	92a-3p	U	A	U	U	G	C	A	C	U	U	G	U	C	C	C	G	G	C	C	U	G	U	

Table 3(B). Nucleotide sequences of suppressive miRNAs for various cancers

cancer entity	has-miR	1	2	3	4	5	6	7	8	9	10	11	12	13	14	15	16	17	18	19	20	21	22	23	24
brain, GBM	128-1-5p	C	G	G	G	G	C	C	G	U	A	G	C	A	C	U	G	U	C	U	G	A	G	A	
	128-1-3p	U	C	A	C	A	G	U	G	A	A	C	C	G	G	U	C	U	C	U	U	U			
	181a-5p	A	A	C	A	U	U	C	A	A	C	G	U	U	G	U	C	G	G	U	G	A	G	U	
	181a-3p	A	C	C	A	U	C	G	A	C	C	G	U	U	G	A	U	U	G	U	A	C	C		
	181b-5p	A	A	C	A	U	U	C	A	U	U	G	C	U	G	U	C	G	G	U	G	G	G	U	
	181b-3p	C	U	C	A	C	U	G	A	A	C	A	A	U	G	A	A	U	G	C	A	A			
	181c-5p	A	A	C	A	U	U	C	A	A	C	C	U	G	U	C	G	G	U	G	A	G	U		
	181c-3p	A	A	C	C	A	U	C	G	A	C	C	G	U	U	G	A	G	U	G	G	A	C		
	181d-5p	A	A	C	A	U	U	C	A	U	U	G	U	U	G	U	C	G	G	U	G	G	G	U	
breast	let-7a-5p	U	G	A	G	G	U	A	G	U	A	G	G	U	U	G	U	A	U	A	G	U	U		
	let-7b-5p	U	G	A	G	G	U	A	G	U	A	G	G	U	U	G	U	G	U	G	G	U	U		
	let-7c-5p	U	G	A	G	G	U	A	G	U	A	G	G	U	U	G	U	A	U	G	G	U	U		
	let-7d-5p	A	G	A	G	G	U	A	G	U	A	G	G	U	U	G	C	A	U	A	G	U	U		
	let-7e-5p	U	G	A	G	G	U	A	G	G	A	G	G	U	U	G	U	A	U	A	G	U	U		
	let-7f-5p	U	G	A	G	G	U	A	G	U	A	G	A	U	U	G	U	A	U	A	G	U	U		
	15a-5p	U	A	G	C	A	G	C	A	C	A	U	A	A	U	G	G	U	U	U	G	U	G		
	16-5p	U	A	G	C	A	G	C	A	C	G	U	A	A	A	U	A	U	U	G	G	C	G		
	125a-5p	U	C	C	C	U	G	A	G	A	C	C	C	U	U	U	A	A	C	C	U	G	U	G	A
	125b-5p	U	C	C	C	U	G	A	G	A	C	C	C	U	A	A	C	U	U	G	U	G	A		
	127-5p	C	U	G	A	A	G	C	U	C	A	G	A	G	G	G	C	U	C	U	G	A	U		
	145-5p	G	U	C	C	A	G	U	U	U	U	C	C	C	A	G	G	A	A	U	C	C	C	U	
	204-5p	U	U	C	C	C	U	U	U	G	U	C	A	U	C	C	U	A	U	G	C	C	U		
lung	let-7a-5p	U	G	A	G	G	U	A	G	U	A	G	G	U	U	G	U	A	U	A	G	U	U		
	let-7b-5p	U	G	A	G	G	U	A	G	U	A	G	G	U	U	G	U	G	U	G	G	U	U		
	let-7c-5p	U	G	A	G	G	U	A	G	U	A	G	G	U	U	G	U	A	U	G	G	U	U		
	let-7d-5p	A	G	A	G	G	U	A	G	U	A	G	G	U	U	G	C	A	U	A	G	U	U		
	let-7e-5p	U	G	A	G	G	U	A	G	G	A	G	G	U	U	G	U	A	U	A	G	U	U		
	let-7f-5p	U	G	A	G	G	U	A	G	U	A	G	A	U	U	G	U	A	U	A	G	U	U		
	9-5p	U	C	U	U	U	G	G	U	U	A	U	C	U	A	G	C	U	G	U	A	U	G	A	
	9-3p	A	U	A	A	A	G	C	U	A	G	A	U	A	A	C	C	G	A	A	A	G	U		
	26a-5p	U	U	C	A	A	G	U	A	A	U	C	C	A	G	G	A	U	A	G	G	C	U		
	27b-5p	A	G	A	G	C	U	U	A	G	C	U	G	A	U	U	G	G	U	G	A	A	C		
	29b-3p	U	A	G	C	A	C	C	A	U	U	U	G	A	A	A	U	C	A	G	U	G	U	U	
	32-5p	U	A	U	U	G	C	A	C	A	U	U	A	C	U	A	A	G	U	U	G	C	A		
	33a-5p	G	U	G	C	A	U	U	G	U	A	G	U	U	G	C	A	U	U	G	C	A			
	30a-5p	U	G	U	A	A	A	C	A	U	C	C	U	C	G	A	C	U	G	G	A	A	G		
	95-3p	U	U	C	A	A	C	G	G	G	U	A	U	U	U	A	U	U	G	A	G	C	A		
	101-3p	U	A	C	A	G	U	A	C	U	G	U	G	A	U	A	A	C	U	G	A	A			
	124-3p	U	A	A	G	G	C	A	C	G	C	G	G	U	G	A	A	U	G	C	C				
	125a-5p	U	C	C	C	U	G	A	G	A	C	C	C	U	U	U	A	A	C	C	U	G	U	G	A
	126-5p	C	A	U	U	A	U	U	A	C	U	U	U	U	G	G	U	A	C	G	C	G			
	126-3p	U	C	G	U	A	C	C	G	U	G	A	G	U	A	A	U	A	A	U	G	C	G		
	140-5p	C	A	G	U	G	G	U	U	U	U	A	C	C	C	U	A	U	G	G	U	A	G		
	140-3p	U	A	C	C	A	C	A	G	G	G	U	A	G	A	A	C	C	A	C	G	G			
	143-3p	U	G	A	G	A	U	G	A	A	G	C	A	C	U	G	U	A	G	C	U	C			
	145-5p	G	U	C	C	A	G	U	U	U	U	C	C	C	A	G	G	A	A	U	C	C	C	U	
	181c-5p	A	A	C	A	U	U	C	A	A	C	C	U	G	U	C	G	G	U	G	A	G	U		
	181c-3p	A	A	C	C	A	U	C	G	A	C	C	G	U	U	G	A	G	U	G	G	A	C		
	198	G	G	U	C	C	A	G	A	G	G	G	G	A	G	A	U	A	G	G	U	U	C		
	192-5p	C	U	G	A	C	C	U	A	U	G	A	A	U	U	G	A	C	A	G	C	C			
	199b-3p	A	C	A	G	U	A	G	U	C	U	G	C	A	C	A	U	U	G	G	U	U	A		
	216a-5p	U	A	A	U	C	U	C	A	G	C	U	G	G	C	A	A	C	U	G	U	G	A		
	216b-5p	A	A	A	U	C	U	C	U	G	C	A	G	G	C	A	A	A	U	G	U	G	A		
	218-5p	U	U	G	U	G	C	U	U	G	A	U	C	U	A	A	C	C	A	U	G	U			
	219a-5p	U	G	A	U	U	G	U	C	C	A	A	A	C	G	C	A	A	U	U	C	U			
	219a-1-3p	A	G	A	G	U	U	G	A	G	U	C	U	G	G	A	C	G	U	C	C	C	G		
	224-5p	C	A	A	G	U	C	A	C	U	A	G	U	G	G	U	U	C	C	G	U	U			

cancer entity	has-miR	1	2	3	4	5	6	7	8	9	10	11	12	13	14	15	16	17	18	19	20	21	22	23	24
esophagus	203a	G	U	G	A	A	A	U	G	U	U	U	A	G	G	A	C	C	A	C	U	A	G		
	205-5p	U	C	C	U	U	C	A	U	U	C	C	A	C	C	G	G	A	G	U	C	U	G		
stomach	let-7a-5p	U	G	A	G	G	U	A	G	U	A	G	G	U	U	G	U	A	U	A	G	U	U		
	let-7b-5p	U	G	A	G	G	U	A	G	U	A	G	G	U	U	G	U	G	U	G	G	U	U		
	let-7c-5p	U	G	A	G	G	U	A	G	U	A	G	G	U	U	G	U	A	U	G	G	U	U		
	let-7d-5p	A	G	A	G	G	U	A	G	U	A	G	G	U	U	G	C	A	U	A	G	U	U		
	let-7e-5p	U	G	A	G	G	U	A	G	G	A	G	G	U	U	G	U	A	U	A	G	U	U		
	let-7f-5p	U	G	A	G	G	U	A	G	U	A	G	A	U	U	G	U	A	U	A	G	U	U		
colorectal	let-7a-5p	U	G	A	G	G	U	A	G	U	A	G	G	U	U	G	U	A	U	A	G	U	U		
	let-7b-5p	U	G	A	G	G	U	A	G	U	A	G	G	U	U	G	U	G	U	G	G	U	U		
	let-7c-5p	U	G	A	G	G	U	A	G	U	A	G	G	U	U	G	U	A	U	G	G	U	U		
	let-7d-5p	A	G	A	G	G	U	A	G	U	A	G	G	U	U	G	C	A	U	A	G	U	U		
	let-7e-5p	U	G	A	G	G	U	A	G	G	A	G	G	U	U	G	U	A	U	A	G	U	U		
	let-7f-5p	U	G	A	G	G	U	A	G	U	A	G	A	U	U	G	U	A	U	A	G	U	U		
	34a-5p	U	G	G	C	A	G	U	G	U	C	U	U	A	G	C	U	G	G	U	U	G	U		
	34b-5p	U	A	G	G	C	A	G	U	G	U	C	A	U	U	A	G	C	U	G	A	U	U	G	
	34b-3p	C	A	A	U	C	A	C	U	A	A	C	U	C	C	A	C	U	G	C	C	A	U		
	34c-5p	A	G	G	C	A	G	U	G	U	A	G	U	U	A	G	C	U	G	A	U	U	G	C	
	127-3p	U	C	G	G	A	U	C	C	G	U	C	U	G	A	G	C	U	U	G	G	C	U		
	133b	U	U	U	G	G	U	C	C	C	C	U	U	C	A	A	C	C	A	G	C	U	A		
	143-3p	U	G	A	G	A	U	G	A	A	G	C	A	C	U	G	U	A	G	C	U	C			
	145-5p	G	U	C	C	A	G	U	U	U	U	C	C	C	A	G	G	A	A	U	C	C	C	U	
hepatic cell	let-7a-5p	U	G	A	G	G	U	A	G	U	A	G	G	U	U	G	U	A	U	A	G	U	U		
	let-7b-5p	U	G	A	G	G	U	A	G	U	A	G	G	U	U	G	U	G	U	G	G	U	U		
	let-7c-5p	U	G	A	G	G	U	A	G	U	A	G	G	U	U	G	U	A	U	G	G	U	U		
	let-7d-5p	A	G	A	G	G	U	A	G	U	A	G	G	U	U	G	C	A	U	A	G	U	U		
	let-7e-5p	U	G	A	G	G	U	A	G	G	A	G	G	U	U	G	U	A	U	A	G	U	U		
	let-7f-5p	U	G	A	G	G	U	A	G	U	A	G	A	U	U	G	U	A	U	A	G	U	U		
	101-3p	U	A	C	A	G	U	A	C	U	G	U	G	A	U	A	A	C	U	G	A	A			
	122-5p	U	G	G	A	G	U	G	U	G	A	C	A	A	U	G	G	U	G	U	U	U	G		
	125a-5p	U	C	C	C	U	G	A	G	A	C	C	C	U	U	U	A	A	C	C	U	G	U	G	A
	195-5p	U	A	G	C	A	G	C	A	C	A	G	A	A	A	U	A	U	U	G	G	C			
	199a-5p	C	C	C	A	G	U	G	U	U	C	A	G	A	C	U	A	C	C	U	G	U	U	C	
	199a-3p	A	C	A	G	U	A	G	U	C	U	G	C	A	C	A	U	U	G	G	U	U	A		
	200a-3p	U	A	A	C	A	C	U	G	U	C	U	G	G	U	A	A	C	G	A	U	G	U		
pancreas	139-5p	U	C	U	A	C	A	G	U	G	C	A	C	G	U	G	U	C	U	C	C	A	G	U	
	142-5p	C	A	U	A	A	A	G	U	A	G	A	A	A	G	C	A	C	U	A	C	U			
	142-3p	U	G	U	A	G	U	G	U	U	U	C	C	U	A	C	U	U	U	A	U	G	G	A	
	345-5p	G	C	U	G	A	C	U	C	C	U	A	G	U	C	C	A	G	G	G	C	U	C		
	375	U	U	U	G	U	U	C	G	U	U	C	G	G	C	U	C	G	C	G	U	G	A		
prostate gland	15a-5p	U	A	G	C	A	G	C	A	C	A	U	A	A	U	G	G	U	U	U	G	U	G		
	16-5p	U	A	G	C	A	G	C	A	C	G	U	A	A	A	U	A	U	U	G	G	C	G		
	143-3p	U	G	A	G	A	U	G	A	A	G	C	A	C	U	G	U	A	G	C	U	C			
	145-5p	G	U	C	C	A	G	U	U	U	U	C	C	C	A	G	G	A	A	U	C	C	C	U	
	218-5p	U	U	G	U	G	C	U	U	G	A	U	C	U	A	A	C	C	A	U	G	U			
cervical	143-3p	U	G	A	G	A	U	G	A	A	G	C	A	C	U	G	U	A	G	C	U	C			
	145-5p	G	U	C	C	A	G	U	U	U	U	C	C	C	A	G	G	A	A	U	C	C	C	U	
B-CELL	15a-5p	U	A	G	C	A	G	C	A	C	A	U	A	A	U	G	G	U	U	U	G	U	G		
	16-5p	U	A	G	C	A	G	C	A	C	G	U	A	A	A	U	A	U	U	G	G	C	G		
	143-3p	U	G	A	G	A	U	G	A	A	G	C	A	C	U	G	U	A	G	C	U	C			
	145-5p	G	U	C	C	A	G	U	U	U	U	C	C	C	A	G	G	A	A	U	C	C	C	U	
	192-5p	C	U	G	A	C	C	U	A	U	G	A	A	U	U	G	A	C	A	G	C	C			
	181a-5p	A	A	C	A	U	U	C	A	A	C	G	U	U	G	U	C	G	G	U	G	A	G	U	
	181a-3p	A	C	C	A	U	C	G	A	C	C	G	U	U	G	A	U	U	G	U	A	C	C		

Table 4(A). Nucleotide frequencies at positions in the oncogenic miRNAs

position	1	2	3	4	5	6	7	8	9	10	11	12	13	14	15	16	17	18	19	20	21	22
A	41	61	46	40	35	21	47	33	24	43	29	39	49	26	34	44	16	36	19	20	15	37
G	1	34	31	20	49	35	37	17	15	21	34	12	23	43	35	38	42	47	43	50	41	27
C	28	26	33	45	11	38	31	34	20	27	36	45	30	27	28	25	31	36	20	17	22	10
U	62	11	22	27	37	38	17	48	73	41	33	36	30	36	35	25	43	13	50	45	54	43

Table 4(B). Nucleotide frequencies at positions in the suppressive miRNAs

position	1	2	3	4	5	6	7	8	9	10	11	12	13	14	15	16	17	18	19	20	21	22
A	22	30	48	24	38	9	41	31	22	46	13	30	21	22	26	26	46	17	26	12	16	9
G	10	45	23	47	42	26	20	47	20	17	46	40	14	21	57	16	20	27	42	54	18	22
C	11	16	29	28	11	17	27	9	16	24	31	22	18	12	10	25	17	10	15	20	26	13
U	68	20	11	12	20	59	23	24	53	24	21	19	58	56	18	44	28	57	28	25	50	46

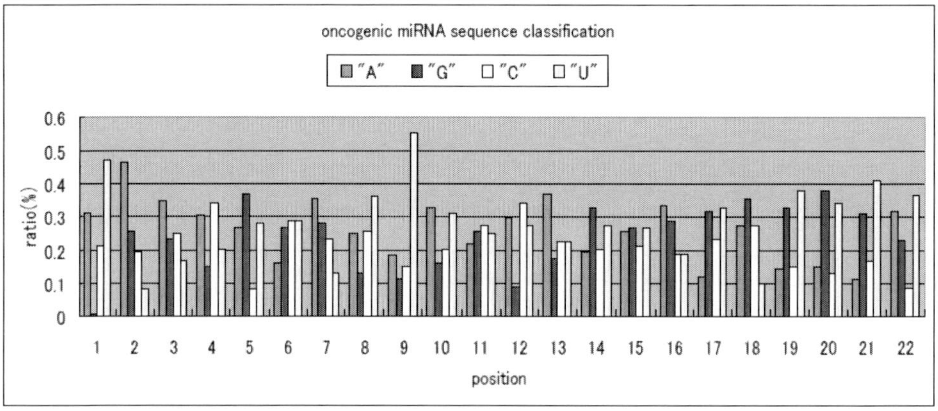

Figure 2(A). Oncogenic miRNA nucleotide distribution ratios.

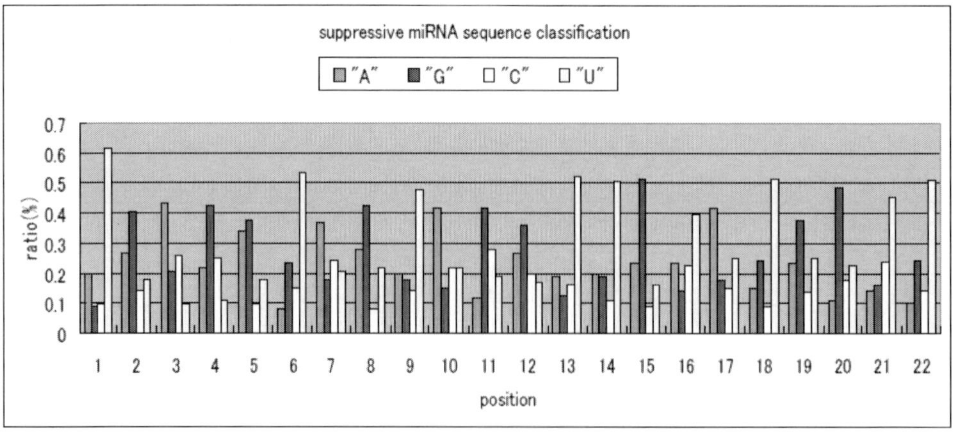

Figure 2(B). Suppressive miRNA nucleotide distribution ratios.

To analyze the differences statistically, the following significance test was carried out.

$$Z_s = \frac{|p_{as} - p_b|}{\sqrt{P(1-P)\left(\frac{1}{n_{as}} + \frac{1}{n_b}\right)}}, \tag{1}$$

where s is the site (position) 1 to 22, p_{as} is the probability of each nucleotide a occurring at each site s (a = A, G, C, or U), p_b is the occurrence probability of each nucleotide averaged over the entire target sequence population, P is the arithmetic mean of p_{as} and p_b, n_{as} is the number of nucleotides at position s, and n_b is the total number of nucleotides in all positions.

As the two-sided statistical test has two types of significance values, higher (upper) and lower levels of significance, they are expressed as follows:

- Higher-significance nucleotide (HN_s^v) and
- Lower-significance nucleotide (LN_s^v),
- where H denotes higher, L denotes lower, and N is a nucleotide,
- v: 95-significance probability is 95% (level of significance = 0.05),
- 99-significance probability is 99% (level of significance = 0.01), and
- s: nucleotide position (site) (i.e., 1–22).

Table 5(A). Values of higher- and lower-significance nucleotides in oncogenic miRNAs

		1	2	3	4	5	6	7	8	9	10	11	12	13	14	15	16	17	18	19	20	21	22
A	*HN*		4.6	2.1				2.3						2.6									
	LN						2.5			2								3.5		3	2.8	3.8	
G	*HN*					3.1									2.2				2.9	2.2	3.3		
	LN	6.1			2.3				2.8	3.3	2		3.9										
C	*HN*				3.1								3.1										
	LN					3.6															2.2		3.3
U	*HN*	4.3								6.1										2.2		2.9	2
	LN		5	2.7	2			3.7									2.3		4.1				

HN: Higher nucleotide. *LN*: Lower nucleotide.

Table 5(B). Values of higher- and lower-significance nucleotides in suppressive miRNAs

		1	2	3	4	5	6	7	8	9	10	11	12	13	14	15	16	17	18	19	20	21	22
A	*HN*			4.2		2.3		3			3.8							3.8					
	LN						3.7					2.7							2		3	2	3.2
G	*HN*		2.8		3	2.2			3			2.8				4.9				2.2	4.5		
	LN	4.2						2.2		2.2	2.9			3.3			3.1	2.2				2.6	
C	*HN*			2.2	2							2.7											
	LN								2.3							2			2.1				
U	*HN*	6.1					4.4			3.4				4.3	3.8				4			2.8	4
	LN		2.9	4.7	5	2.9		2.3	2.1		2.1	2.7	3.2			3.4							

HN: Higher nucleotide. *LN*: Lower nucleotide.

miRNA GENE-SILENCING SCORE

Nucleotide frequencies at the individual positions listed in Tables 4(A) and 4(B) were analyzed by using Eq. (1), and many higher-significance and lower-significance nucleotides were obtained. They are listed in Tables 5(A) and 5(B). A miRNA sequence including many higher-significance nucleotides and a few lower-significance nucleotides could be inferred to be a highly effective gene silencer. A miRNA gene-silencing score S based on this idea is therefore defined as the sum of the higher-significance nucleotides in the sequence minus the sum of the lower-significance nucleotides in the sequence. That is,

$$S = \sum HN_s^v - \sum LN_s^v \,, \tag{2}$$

where HN_s^v and LN_s^v are higher-significance and lower-significance nucleotides, respectively. The larger S is, the greater the likelihood that the sequence is effective for gene silencing. miRNA scores for the oncogenic and suppressive miRNAs listed in Tables 3(A) and 3(B) were calculated using Eq. (2). Their sums and averages for the individual cancers are listed in Table 6. One sees in Table 6 and Figure 3 that overall there is a tendency that the average scores of suppressive miRNAs are greater than those of oncogenic ones. This dependency may be due to the big role that let-7 miRNAs play as suppressive miRNAs.

Table 6. Sums and averages of oncogenic and suppressive miRNAs for various cancers

	oncogenenic miRNAs		suppresive miRNAs	
	sum	average	sum	average
GBM	37.37	12.46	79.86	8.87
breast	126.4	6.32	529.8	40.75
lung	168.3	8.86	633.1	17.59
esophagus	43.83	21.92	-0.35	-0.165
stomach	112.2	9.35	411.37	68.56
colorectal	282.1	14.85	520.7	37.20
hepatic cell	122.8	13.65	528.4	40.65
pancreas	354.6	11.44	-14.05	-2.81
prostage gland	255.2	18.23	80.45	16.1
uterus	37.8	18.9	35.58	17.79
B-CELL	32.98	10.93	83.3	11.9
Total	1573	11.74	2888.2	25.12

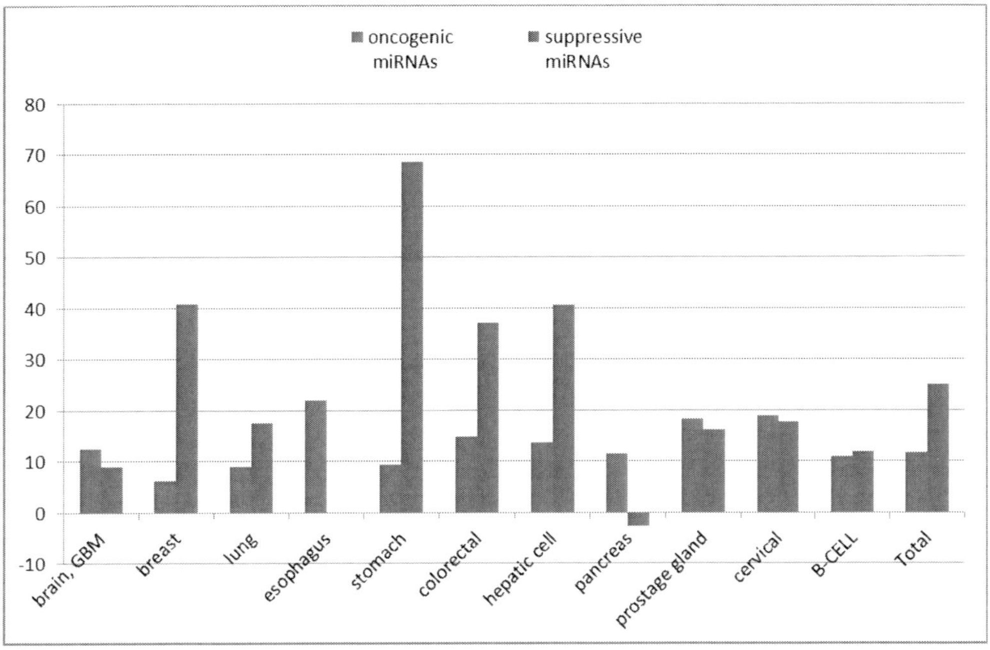

Figure 3. Relations between average scores of oncogenic and suppressive miRNAs for various cancers.

HIGHER- AND LOWER-LEVEL NUCLEOTIDES OF ONCOGENIC AND SUPPRESSIVE miRNAS

Higher- and lower-level nucleotides at individual positions in oncogenic and suppressive miRNAs are listed in Tables 7(A) and 7(B). Since the higher-level nucleotides at individual positions are those that have a larger influence on the upregulation or downregulation due to the miRNAs, the miRNA sequences including many higher-level nucleotides may have a big impact on the target mRNA silencing. With this in mind, the coincidences between highly frequent miRNA sequences and higher-level nucleotides were examined for oncogenic and suppressive miRNAs of individual cancers. The results are listed in Tables 8(A) and 8(B). Ten of the kinds of cancers listed in Table 2(A) are stimulated by hsa-miR-21, but one sees in Table 8(A) that there are only six nucleotide coincidences between the miR-21 sequence and higher-level nucleotides. On the other hand, although only six kinds of cancers are stimulated by hsa-miR-17 and only five kinds are stimulated by has-mir-191, the numbers of nucleotide coincidences between miR-17 and -191 sequences and higher-level nucleotides are respectively ten and nine. This implies that the miR-17 and -191 sequences influence the target mRNA regulations more than the miR-21 sequence does. In contrast, hsa-miR-145 suppresses six kinds of cancers and has seven coincident nucleotides listed in Table 8(B), whereas although both let-7 and hsa-miR-143 suppress five kinds of cancers, their nucleotide coincidences are respectively nineteen and ten as listed in Table 8(B). This also implies that the influences of let-7 and miR-143 are greater than the influence of miR-145.

Table 7(A). Higher- and lower-level nucleotides in oncogenic miRNAs

position	1	2	3	4	5	6	7	8	9	10	11	12	13	14	15	16	17	18	19	20	21	22
Higher	U	A	A	C	G	X	A	X	U	X	X	C	A	G	X	X	X	G	G	G	U	U
																			U			
Lower	G	U	U	G	C	A	U	G	G	G	X	G	X	X	X	U	A	U	A	A	A	C
				U					A											C		

X: not specified nucleotide, i.e., X=A, G, C, or U.

Table 7(B). Higher- and lower-level nucleotides in suppressive miRNAs

position	1	2	3	4	5	6	7	8	9	10	11	12	13	14	15	16	17	18	19	20	21	22
Higher	U	G	A	G	A	U	A	G	U	A	G	X	U	U	G	X	A	U	G	G	U	U
			C	C	G						C											
Lower	G	U	U	U	U	A	G	C	G	G	A	U	G	X	U	G	G	A	X	A	C	A
							U	U		U	U				C			C			A	

X: not specified nucleotide, i.e., X=A, G, C, or U.

Table 8(A). Relations between higher-level nucleotides and highly frequent oncogenic miRNAs

position	1	2	3	4	5	6	7	8	9	10	11	12	13	14	15	16	17	18	19	20	21	22
Higher nucleotides	U	A	A	C	G	X	A	X	U	X	X	C	A	G	X	X	X	G	G	G	U	U
																			U			
miR-21-5p	U	A	G	C	U	U	A	U	C	A	G	A	C	U	G	A	U	G	U	U	G	A
	O	O		O			O											O	O			
miR-17-5p	C	A	A	A	G	U	G	C	U	U	A	C	A	G	U	G	C	A	G	G	U	A
		O	O		O				O			O	O	O					O	O	O	
miR-191-5p	C	A	A	C	G	G	A	A	U	C	C	C	A	A	A	A	G	C	A	G	C	U
		O	O		O		O		O			O	O							O		O
miR-221-5p	A	C	C	U	G	G	C	A	U	A	C	A	A	U	G	U	A	G	A	U	U	U
					O				O				O					O			O	O

X: not specified nucleotide, i.e., X=A, G, C, or U.

Table 8(B). Relations between higher-level nucleotides and highly frequent suppressive miRNAs

position	1	2	3	4	5	6	7	8	9	10	11	12	13	14	15	16	17	18	19	20	21	22
Higher nucleotides	U	G	A	G	A	U	A	G	U	A	G	X	U	U	G	X	A	U	G	G	U	U
			C	C	G						C											
miR-145-5p	G	U	C	C	A	G	U	U	U	U	C	C	C	A	G	G	A	A	U	C	C	C
			O	O	O				O		O				O		O					
let-7e-5p	U	G	A	G	G	U	A	G	U	A	G	G	U	U	G	U	A	U	G	G	U	U
	O	O	O	O		O	O	O	O	O	O		O	O	O		O	O	O	O	O	O
miR-143-3p	U	G	A	G	A	U	G	A	A	G	C	A	C	U	G	U	A	G	C	U	C	
	O	O	O	O	O	O					O			O	O		O					

X: not specified nucleotide, i.e., X=A, G, C, or U.

RELATIONS BETWEEN miRNA SCORES AND NUCLEOTIDE COINCIDENCES

Then the relations between miRNA scores and the number of coincident nucleotides of oncogenic and suppressive miRNA sequences were examined. As miRNA scores indicated different ranges from low to high for individual miRNA sequences, the score ranges were used to examine the relations between the scores and nucleotide coincidences of the higher and lower significance levels.

Table 9(A). Relations between miRNA scores and the number of coincident nucleotides in oncogenic miRNAs

score range	Higher	Lower	Total
31 ~ 35	10	0	10
26 ~ 30	10.1	1.33	8.8
21 ~ 25	7.25	0.25	7
16 ~ 20	6.8	1.5	5.3
11 ~ 15	6.9	3.2	3.7
6 ~ 10	5.7	3	2.3
1 ~ 5	4.9	4.37	0.5
-5 ~ 0	3.5	4.27	-0.8
-10 ~ -6	3.2	5.33	-2.1
-16 ~ -11	1	5.5	-4.5

Higher: No. of average coincident nucleotides at higher level.
Lower: No. of average coincident nucleotides at lower level. Total: Higher – Lower.

Table 9(B). Relations between miRNA scores and the number of coincident nucleotides in suppressive miRNAs

score range	Higher	Lower	Total
71 ~	20	0	20
66 ~ 70	19	0.33	18.7
61 ~ 65	18	0.5	17.5
31 ~ 35	11.3	2.7	8.6
26 ~ 30	9.25	2	7.25
21 ~ 25	9.63	3.75	5.88
16 ~ 20	8.83	3.83	5
11 ~ 15	7.78	4.78	3
6 ~ 10	7.11	5.11	2
1 ~ 5	6.38	6.76	-0.38
-5 ~ 0	4.92	5.83	-0.92
-10 ~ -6	5	8	-3
-16 ~ -11	3.25	8	-4.75

Higher: No. of average coincident nucleotides at higher level.
Lower: No. of average coincident nucleotides at lower level. Total: Higher – Lower.

In addition, the average coincident nucleotides in the score ranges were used at the higher and lower levels. Total coincident nucleotides were calculated as the number of coincident nucleotides at the higher level minus the number of coincident nucleotides at the lower level. The relations between miRNA scores and the number of coincident nucleotides of oncogenic and suppressive miRNAs are shown in Tables 9(A) and 9(B) and in Figs. 4(A) and 4(B). It is clear that the number of coincident nucleotides at the higher level for oncogenic miRNAs decreases gradually as the score ranges decrease, whereas the number of coincident nucleotides at the lower level increases as the score ranges decrease; and the total number of coincident nucleotides decreases distinctly as shown in Figure 4(A). This tendency is similar to the tendency in Figure 4(B).

Comparing Table 7(A) with Table 7(B), it become clear what nucleotides occur coincidentally in oncogenic and suppressive miRNA sequences at higher and lower significance levels.

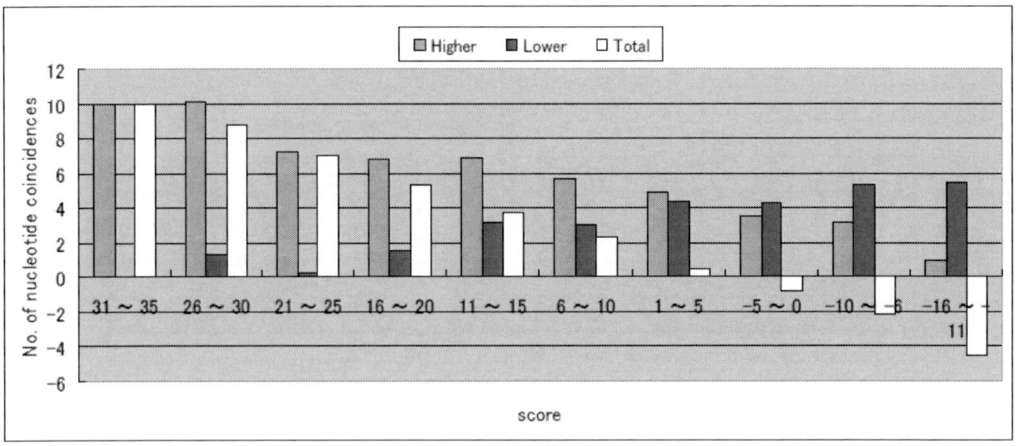

Figure 4(A). Relations between miRNA scores and the number of coincident nucleotides in oncogenic miRNAs.

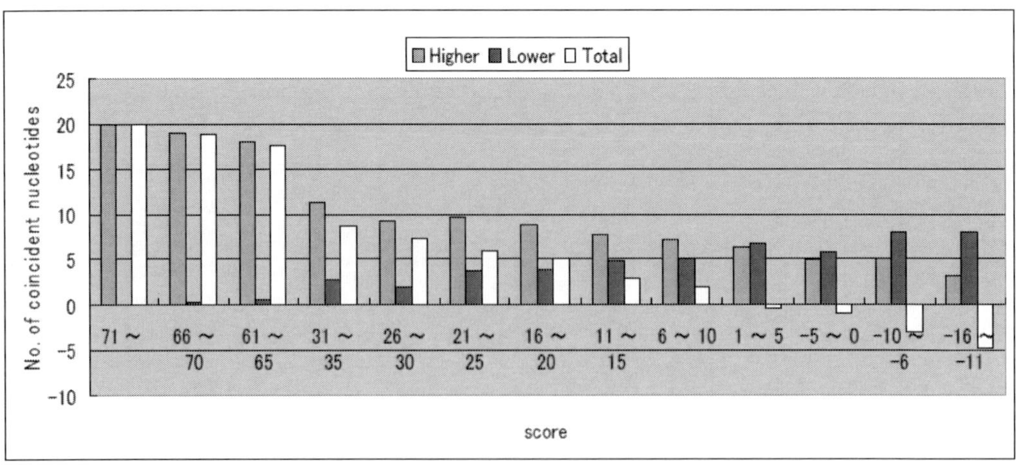

Figure 4(B). Relations between miRNA scores and the number of coincident nucleotides in suppressive miRNAs.

In case of the higher level, the nucleotides U, A, C, G, A, U, G, G, U, and U respectively occurred at positions 1, 3, 4, 5, 7, 9, 19, 20, 21, and 22 in both oncogenic and suppressive miRNAs. On the other hand, in case of the lower level, the nucleotides G, U, U, U, A, U, G, G, A, and A respectively occurred at positions 1, 2, 3, 4, 6, 7, 9, 10, 20, and 21. These nucleotide coincidences imply that it is difficult to tell from only the higher- and lower-level nucleotides whether a miRNA is an oncogenic or suppressive one.

FREQUENCIES OF INDIVIDUAL NUCLEOTIDES IN miRNAS

Frequencies of individual nucleotides were also examined for oncogenic and suppressive miRNAs in both the higher and lower significances as shown in Figure 5. It is clear that the most significant nucleotide as a whole is U. At the lower level it plays an especially important role in suppressive miRNAs. G is also often a frequently occurring nucleotide. At the lower level, A occurs with almost the same frequency as G. The least frequently occurring nucleotide is C. The occurrences of U at positions 2 to 12 in the lower-level rows of Table 7(B) indicate that this nucleotide has strong influences on the suppressive miRNAs.

It has been believed that nucleotides 2 to 8 in the 5' portion of miRNAs are most important for target recognition by vertebrate miRNAs. The higher- and lower-level nucleotides in oncogenic and suppressive miRNAs listed in Tables 7(A) and 7(B) were examined from that point of view. The nucleotide coincidences of the higher levels between oncogenic and suppressive miRNAs are four (i.e., at positions 3 – A, 4 – C, 5 – G, and 7 – A), whereas those of the lower level are five (at positions 2 – U, 3 – U, 4 – U, 6 – A, and 7 – U). At the higher level the most frequently occurring nucleotide is A (six) and the second-most frequently occurring one is G (five), whereas at the lower level the most frequently occurring nucleotide is U (ten) and the second-most frequently occurring one is G (three).

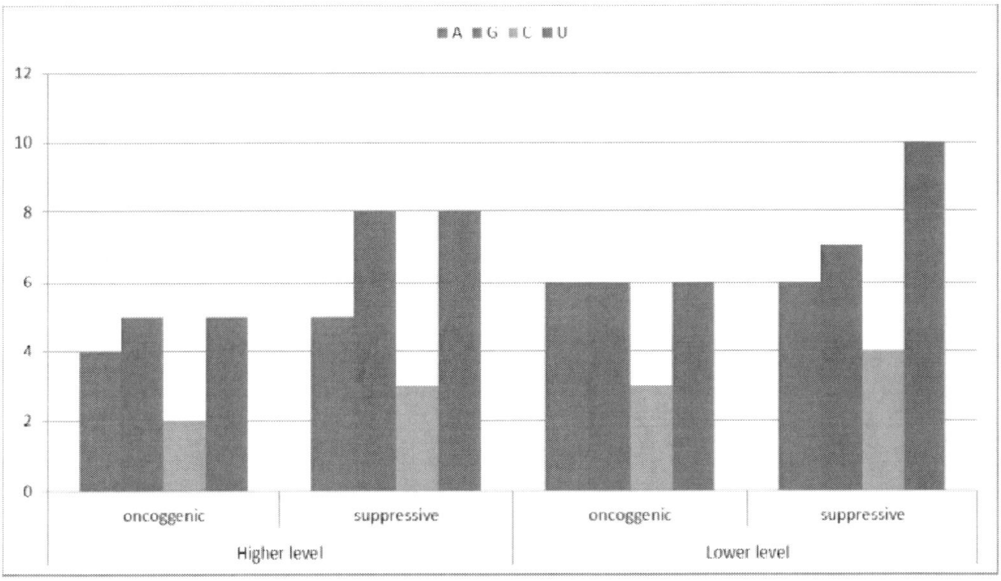

Figure 5. Frequencies of individual nucleotides at higher and lower levels.

DISTINGUISHING WHETHER A NEW miRNA IS ONCOGENIC OR SUPPRESSIVE

It is important to know what function a new miRNA has, especially whether the miRNA is oncogenic or suppressive. A new method to distinguish the function of a new miRNA can be based on the higher-level nucleotides and miRNA scores and is shown in Figure 6. The determination processes are as follows:

Step 1: Examine the possibility that it is an oncogenic miRNA.

- Count the numbers of higher- and lower-level nucleotides in it according to the nucleotides listed in Table 7(A).
- Calculate the score of the new miRNA by using Eq. (2) and Table 5(A).

Step 2: Examine the possibility that it is a suppressive miRNA.

- Count the numbers of higher- and lower-level nucleotides in it according to the nucleotides listed in Table 7(B).
- Calculate the score of the new miRNA by using Eq. (2) and Table 5(B).

Step 3: Distinguish whether the new miRNA is an oncogenic one or a suppressive one.

- Whether the new miRNA is oncogenic or suppressive is determined on the basis of Figs. 4(A) and 4(B) and Tables 7(A) and 7(B). If the numbers of the higher- and lower-level nucleotides of the new miRNA are respectively greater than ten and less than two (i.e., the total number is greater than eight) and the score based on Figure 4(A) and Table 7(A) is greater than 26, the new miRNA is likely to be oncogenic. If, on the other hand, the total number is greater than 17 and the score based on Figure 4(B) and Table 7(B) is greater than 61, the new miRNA is likely to be suppressive. If neither of these conditions applies to the new miRNA, the possibility of cancer regulations by the new miRNA is estimated to be low.

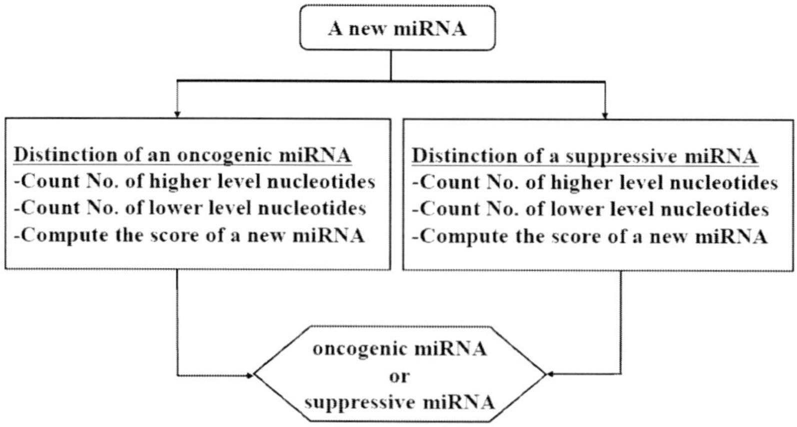

Figure 6. Distinguishing whether a new miRNA is oncogenic or suppressive.

Conclusion

In this article, the roles of miRNAs in human cancers are examined from the viewpoint of up- and down-regulation. As a result, the statistical analysis of the positional nucleotide occurrence features of miRNAs revealed differences between the positional nucleotide occurrences of oncogenic and suppressive miRNAs. A miRNA gene-silencing score was then defined on the basis of the higher and lower levels of the statistical significances of positional nucleotides. Since the miRNA scores were closely related to miRNA frequencies, a method using the scores and nucleotide frequencies to distinguish whether a new miRNA is oncogenic or suppressive was proposed.

References

[1] Aravin A, Logos-Quintana M, Yalcin A, Zavolan M, Marks D, Snyder B, Gaasterland T, Meyer J, Tuschl T. (2003) The small RNA profile during *Drosophila melanogaster* development. *Developmental Cell* 5: 337–50.

[2] Bartel DP. (2004) MicroRNAs: genomics, biogenesis, mechanism, and function. *Cell* 116: 281–97.

[3] Doench JG, Sharp PA. (2004) Specificity of microRNA target selection in translational repression. *Genes & Development* 18: 504–11.

[4] Lai EC. (2002) Micro RNAs are complementary to 3' UTR sequence motifs that mediate negative post-transcriptional regulation. *Nature Genetics* 30: 363–4.

[5] Volinia S, Calin GA, Liu CG et al. (2006) A microRNA expression signature of human solid tumors defines cancer gene targets. *Proceedings of the National Academy of Sciences USA* 103: 2257–2261.

[6] Chan JA, Krichevsky AM, Kosik KS. (2005) MicroRNA-21 is an antiapoptotic factor inhuman glioblastoma cells. *Cancer Research* 65: 6029–6033.

[7] Takeshita F, MinaY, Nagahara S et al. (2005) Efficient delivery of small interfering RNA to bone-metastatic tumors by using atelocollagen in vivo. *Proceedings of the National Academy of Sciences USA* 102: 12177–12182.

[8] Calin GA, Ferracin M, Cimmino A et al. (2005) A microRNA signature associated with prognosis and progression in chronic lymphocytic leukemia. *New England Journal of Medicine* 353: 1793–1801.

[9] Saito Y, Liang G, Egger G et al. (2006) Specific activation of microRNA-127 with downregulation of the proto-oncogene BCL-6 by chromatin-modifying drugs in human cancer cells. *Cancer Cell* 9: 435–443.

[10] Takamizawa J, Konishi H, Yanagisawa K et al. (2004) Reduced expression of the let-7 microRNAs in human lung cancers in association with shortened postoperative survival. *Cancer Research* 64: 3753–3756.

[11] Zhang HH, Wang XJ, Li GX et al. (2007) Deletion of let-7a microRNA by real time PCR in gastric carcinoma. *World Journal of Gastroenterology* 13: 2883–2888.

[12] Yanaihara N, Caplen N, Bowman E et al. (2006) Unique microRNA molecular profiles in lung cancer diagnosis and prognosis. *Cancer Cell* 9: 189–198.

[13] Feber A, Xi L, Luketich JD et al. (2008) MicroRNA expression profiles of esophageal cancer. *Journal of Thoracic and Cardiovascular Surgery* 135: 255–260.

[14] Bandres E, Cubedo E, Agirre X et al. (2006) Identification by real-time PCR of 13 mature microRNAs differentially expressed in colorectal cancer and non-tumoral tissues. *Molecular Cancer* 5: 29.

[15] Gramantieri L, Ferracin M, Fornari F et al. (2007) Cyclin G1 is a target of miR-122a, a microRNA frequency down-regulated in human hepatocellular carcinoma. *Cancer Research* 67: 6092–6099.

[16] MZ, O'Conner SM, van Holst Pellekaan NG et al. (2003) Reduced accumulation of specific microRNAs in colorectal neoplasia. *Molecular Cancer Research* 1: 882–891.

[17] Meng F, Henson R, Wehbe-Janek H et al. (2007) MicroRNA-21 regulates expression of the PTEN tumor suppressor gene in human hepatocellular cancer. *Gastroenterology* 133: 647–658.

[18] Murakami Y, Yasuda T, Saigo K et al. (2006) Comprehensive analysis of microRNA expression patterns in hepatocellular carcinoma and non-tumorous tissues. *Oncogene* 25: 2537–2545.

[19] Esquela-kerscher A, Slack FJ. (2006) Oncomirs—microRNAs with a role in cancer. *Nature Reviews Cancer* 6: 259–269.

[20] Lui WO, Pourmand N, Patterson BK, Fire A. (2007) Patterns of known and novel small RNAs in human cervical cancer. *Cancer Research* 67: 6031–6043.

[21] Akao Y, Nakagawa Y, Kitade Y et al. (2007) Downregulation of microRNAs-143 and -145 in B-cell malignancies. *Cancer Science* 98: 1914–1920.

[22] Calin GA, Dumitru CD, Shimizu M et al. (2002) Frequent deletions and down-regulation of micro-RNA genes miR15 and miR16 at 13q14 in chronic lymphocytic leukemia. *Proceedings of the National Academy of Sciences* USA 99: 15524–15529.

[23] Calin GA, Croce CM. (2006) MicroRNA signatures in human cancers. *Nature Reviews Cancer 6:* 857–866.

[24] Calin GA, Sevignani C, Dumitru CD et al. (2004) Human microRNA genes are frequently located at fragile sites and genomic regions involved in cancers. *Proceedings of the National Academy of Sciences USA* 101: 2999–3004.

[25] Calin CA, Liu CG, Sevignani C et al. (2004) MicroRNA profiling reveals distinct signatures in B cell chronic lymphocytic leukemias. *Proceedings of the National Academy of Sciences USA* 101: 11755–11760.

[26] Osaki M, Takeshita F, Ochiya T. (2008) MicroRNAs as biomarkers and therapeutic drugs in human cancer. *Biomarkers* 13: 658–670.

[27] Lawrie CH, Gal S, Dunlop HM et al. (2008) Detection of clevated levels of tumour-associated microRNAs in serum of patients with diffuse large B-cell lymphoma. *Br. J. Haematol.* 141: 672 – 675.

[28] Mitchell PS, Parkin RK, Kroh EM et al. (2008) Circulating microRNAs as stable blood-based markers for cancer detection. *Proc. Natl. Acad. Aci. USA* 105: 10513 – 10518.

[29] Taylor DD, Gercel-Taylor C. (2008) MicroRNA signatures of tumor-derived exosomes as diagnostic biomarkers of ovarian cancer. *Gymecol. Oncol.* 110: 13 – 21.

[30] Resnick KE, Alder H, Hagan JP et al. (2009) The detection of differentially expressed microRNAs from the serum of ovarian cancer patients using a novel real-time PCR platform. *Gynecol. Oncol.* 112: 55 – 59.

[31] Hu Z, Chen X, Zhao Y et al. (2010) Serum microRNA signatures identified in a genome-wide serum microRNA expression profiling predict survival of non-small-cell lung cancer. *J. Clin. Oncol.* 28: 1721 – 1726.

[32] Tanaka M, Oikawa K, Takanashi M et al. (2009) Down-regulation of miR-92 in human plasma is a novel marker for acute leukemia patients. *PloS ONE* 4: e5532.

[33] Heneghan HM, Miller N, Lowery AJ et al. (2010) Circulating microRNAs as novel minimally invasive biomarkers for breast cancer. *Ann. Sur.* 251: 499 – 505.

[34] Zhu W, Qin W, Atasory U, Sauter ER. (2009) Circulating microRNAs in breast cancer and healthy subjects. *BMC Res. Notes* 2: 89.

[35] Tsujiura M, Ichikawa D, Komatsu S et al. (2010) Circulating microRNAs in plasma of patients with gastric cancers. *Br. J. Cancer* 102: 1174 – 1179.

[36] Ho AS, Huang X, Cao H et al. (2010) Circulating miR-210 as a novel hypoxia marker in pancreatic cancer. *Transl. Oncol.* 3: 109 – 113.

[37] Wang J, Chen J, Chang P et al. (2009) MicroRNAs inplasma of pancreatic ductal adenocarcinoma patients as novel blood-based biomarkers of disease. *Cancer Prev. Res.* (Phila Pa) 2: 807 – 813.

[38] Wong TS, Liu XB, Wong BY et al. (2008) Mature miR-184 as potential oncogenic microRNA of squamous cell carcinoma of tongue. *Clin. Cancer Res.* 14: 2588 – 2592.

[39] Ng EK, Chong WW, Jin H et al. (2009) Differential expression of microRNAs in plasma of patients with colorectal cancer: a potential marker for colorectal cancer screening. *Gut* 58: 1375 – 1381.

[40] Yamamoto Y, Kosaka N, Tanaka M et al. (2009) MicroRNA-500 as a potential diagnostic marker for hepatocellular carcinoma. *Biomarkers* 14: 529 – 538.

[41] Kosaka N, Iguchi H, Ochiya T. (2010) Circulating microRNA in body fluid: a new potential biomarker for cancer diagnosis and prognosis. *Cancer Sci.* 101: 2087 – 2092.

[42] Griffiths-Jones S, Grocock RJ, van Dongen S, Bateman A, Enright AJ. (2006) miRBase: microRNA sequences, targets and gene nomenclature. *Nucleic Acids Research* 34: D140–4.

[43] Kozomara, A, Griffiths-Jones S, (2010) miRBase: integrating microRNA annotation and deep-sequencing data. *Nucleic Acids Research* 39: D152–7.

In: Advances in Genetics Research. Volume 12
Editor: Kevin V. Urbano
ISBN: 978-1-63321-351-7

Chapter 8

GENETIC PROGRAMMING FOR MUSIC

Maximos Kaliakatsos–Papakostas[1,*] ***Andreas Floros***[2]
and Michael N. Vrahatis[3]
[1]School of Music Studies, Aristotle University of Thessaloniki,
Thessaloniki, Greece
[2]Department of Audio and Visual Arts,
Ionian University, Corfu, Greece
[3]Department of Mathematics,
University of Patras, Patras, Greece

Keywords: music, genetic programming, evolutionary music composition, evolutionary sound synthesis, musical audio analysis

Abstract

Analyzing and generating music material has been a field of intense research, combining scientific disciplines from computational intelligence and signal processing or computational music analysis. This chapter reviews the application of methods that utilize genetic programming (GP) along with other methods for tackling several tasks that pertain to audio and music analysis and synthesis. The focus of this overview is in the evolution of the methodologies themselves and therefore, the discussed papers are mostly presented in a chronological order. In musical audio analysis, GP–based methodologies were developed to performed musical instrument identification, i.e. to recognize the instruments that are included in musical recordings. For audio synthesis, GP has been primarily utilized to evolve combinations of sound generating blocks or simple mathematical functions that shape waveforms. Symbolic music composition has been mostly approach by evolving simple programs that modify short melodic segments and concatenate them to compose a complete melodic part. For the sound and music synthesis methodologies, the main concern revolves around formalizing efficient evaluation techniques for the computation of the fitness value. On the one hand, interactive evaluation, where the evolution is driven through user ratings, incorporates the phenomenon of user fatigue, according to which the users provide small numbers of potentially contradictory evaluations. On the other hand, efficient automatic evaluation requires the development of qualitative music and sound descriptors, a fact that

*E-mail address: maxk@mus.auth.gr

constitutes by itself a deep research subject. The chapter concludes by addressing issues concerning the utilization of GP towards approaching effective solutions to define proper music and sound features, as well as to alleviate user fatigue.

1. Introduction

The creation of artistic material through automated means represents a broad field of study that from a technological point of view basically aims to relate abstract human thought with algorithmic processes. This chapter discusses the application of algorithmic frameworks built on the basis of genetic programming (GP) that are related to musical tasks. In these frameworks the power of GP is harnessed with other techniques, e.g. signal processing or computational music analysis, to produce or analyze music content. Under this scope, this chapter provides an overview of methodologies that utilize GP for musical *sound analysis*, *sound synthesis* and *automatic music composition*. The majority of the papers discussed in this review are commented in considerable detail. Aim of this review however is not only to document the contribution of the GP community to musical analysis and synthesis, but also to highlight the flow of research in this field. Therefore, a historical priority has been given to the discussed subjects, focussing on two important and reciprocal aspects: innovation in the algorithmic design and efficiency measurements of the applied methodologies. Additionally, it is among the main goals of this review to outline the utilization of combined methodologies from various fields with GP, as it is indispensable for all the studies discussing this highly multidisciplinary topic.

Regarding sound analysis, GP methodologies have focussed on combining signal processing techniques to achieve instrument identifications from recordings. This identification is accomplished by labeling a recorded audio part according to the instrument(s) it includes. The utilization of GP in this area incorporated the evolution of regression functions that categorize audio features, extracted through signal processing techniques, providing labels for the music instrument that they correspond to. For sound synthesis, the main research goal revolved around the construction of sound generators through GP evolution. The latter techniques produced sounds that were similar to a target sound – typically an acoustic musical instrument. Fitness evaluation was performed through signal processing techniques, by comparing several spectral characteristics of the produced signals with respective characteristics of the targeted sounds. Interactive evolution has also been examined as a case, but the implications of *user fatigue*, in combination with the fact that the examined sound generators often produced noisy output, did not allow for the achievement of encouraging results. Regarding sound analysis and synthesis, the reported results were mostly promising, as partially confirmed by subjective tests assessing the similarity between target and artificial instrumental sounds.

For automated music composition, the application of GP has mainly focussed on evolving combinations of simple programs that received simple musical phrases as arguments and exported altered versions of these phrases, regarding both pitches and rhythm. Fitness evaluation in evolutionary music composition is among the most fertile fields in the scientific conversation. On the one hand, automated music generation systems that are adaptive (like the ones utilizing evolutionary techniques) have two extreme states regarding fitness: on the one extreme, a very strict definition of the desired output will force an effective system to adapt exactly on the targeted melody, reproducing exact musical "clones", while on the other extreme, the music it composes would be completely random. On the other hand, the definition of qualitative music fitness criteria is by itself a very difficult subject. Therefore, the utilization of GP for music composition has focused on either developing efficient qualitative criteria for music (for non–interactive methods) or on alleviating user fatigue (for interactive methods). The problem of developing proper fitness criteria has also led to the employment of advanced machine learning techniques, e.g. artificial neural networks and self organizing maps. Interactive methodologies have examined several techniques to alleviate user fatigue, as conducting an intermediate automatic evaluation–rejection of poorly fit individuals, or providing the user with the ability to directly change the phenotype of an individual, or even incorporating the employment of fitness–adaptive GP genetic operators.

For this book chapter, the reader is assumed to be acquainted with the basic definitions and notions of GP. For clarifying the utilized terminology, a GP individual is considered to be expressed by a tree representation of programs or functions called the *genotype*. The interpretation of the genotype to the unit that produces the required output (e.g. a melody) is called the *phenotype* of the individual. The materialization of an individual into an output unit is what defines its suitability to comprise a solution to the problem at hand and therefore, the *fitness evaluation* of an individual is performed over the phenotype, while the numeric value of the fitness expresses its suitability. The GP individuals are divided in different generations, with *children* or *offspring* being generated by *parents* through the utilization of GP operators (like crossover and mutation). Each individual is composed of *terminal* and *non–terminal* nodes. Terminal nodes can be though of as the leaves in the tree representation of an individual, i.e. nodes that are only children, while non–terminal nodes can be both children and parents (the utilization of the terms "children" and "parents" in this sentence are loaned from graph theory and should not be confused the respective GP terms).

2. Musical Audio Analysis and Sound Synthesis

Automated analysis of music signal has been widely pursed for several tasks, since its prospects lead to the extraction of descriptive information for large datasets of audio material without laborious human intervention. The GP–related methodologies developed so far aim to analyze musical audio recordings, providing information about which musical instruments are incorporated. Automated sound synthesis on the other hand, is focussed towards yielding systems that generate sound that resembles the sound produced by acoustic instruments. To this end, GP either evolves mathematical expressions that directly produce waveforms, or evolve combinations of sound production, filtering and modulation blocks, according to the fitness evaluation provided by the spectral characteristics of the generated and the targeted sounds. Results on both fields indicate that GP is a promising technique that should be further studied.

2.1. Musical Audio Analysis

The field of study in audio analysis with GP methodologies hitherto concerns the musical instrument identification. This task discusses the labeling of recorded music parts, according to the instruments that they incorporate. The first step in methodologies that perform instrument identification (or source identification, in general) is to extract a set of audio features from the recordings under examination. These audio features are subsequently used for categorizing the recorded excerpts, by utilizing, more often, supervised machine learning techniques, e.g. multi–layer perceptrons [23], *k*-th nearest neighbors [8] and support vector machines [10] among others [16]. These studies may include a great number of features for the classification of sounds. Among the more often used features are those that are related with the human perception of sound, like the mel–frequency cepstral coefficients, or various others that include spectral information (e.g. spectral centroid, roll–off or derivative). The number of instruments to be classified ranges from 2 [5] to 30 [9].

An advanced strategy regarding audio signal processing for instrument identification using GP, has been studied in [24]. The approach investigated therein discusses labeling of audio data that incorporate multiple instruments. Although a thorough presentation of the audio engineering aspect of the aforementioned paper is beyond the scope of this chapter, a brief and intuitive description of the audio features that have been employed is necessary for approaching the immense power that GP allows to methodologies that manipulate great quantities of features. The first step in the study under discussion was to extract many audio features from manually labeled musical recordings. These features are computed for the "background" and the "foreground" of the examined recordings by applying a source separation algorithm. The separation of the audio material in foreground and background leads to a first clarification of the audio material under examination, providing two waveforms which incorporate musical instruments with different spectral characteristics.

Afterwards, a set of 51 basic functions are combined with GP. The resulting combinations are also functions themselves which receive input from the features that are extracted from the foreground and background parts of the recording, while their output is a numeric value that indicates the existence of a specific music instrument. The reported results concern the extraction of several audio features from above 21000 1-second long musical recordings, which had been previously labeled by the authors. These audio recordings were actually 1-second slices from continuous parts of popular commercial songs and have been labeled according to which instruments (among a total set of 10) they include. The mean reported accuracy of the instrument identification process was around 88%, with maximum and minimum values ranging from 76% to 96% (depending on the instrument identified).

A different approach to the problem of musical instrument recognition was examined in Chapter 7 of the PhD thesis in [27]. An extension of the work presented therein, which will be analyzed in the context of the book chapter at hand, was recently presented in [26]. Instrument recognition in the work under discussion is defined as a different task than the one presented in the previous paragraph. Specifically, while in the previously examined work, a musical recording was labeled according to the – potentially many – instruments it included, the latter approach investigated audio clips that include a single musical instrument and answers to the question of which – single – instrument is present in a recording. Therefore, there is no need to separate the waveform under examination to foreground and background, as executed in the previously discussed work.

For the single–instrument recognition presented in [26], a similar approach to the one presented for multi–instrument identification was followed. First, several audio features were extracted from the sound clips. These features included various temporal and spectral features that describe statistical properties of the recordings' spectrogram. Thereafter, the extracted features were given as input to simple mathematical functions – named as fitness methods – that produced a single output, which determined the instrument identified by this function. To highlight the effectiveness and simplicity at the same time of this GP approach, the type of the mathematical fitness methods utilized are discussed in further detail, by categorizing them in the following two groups:

1. *Single output* methods: mathematical expression that incorporated arithmetic operators (e.g. addition, subtraction, trigonometric functions, etc.) and boolean logic (e.g. OR, AND, NOT, etc.). Their output is a single integer value, with 1 denoting piano, 2 violin, 3 flute, 4 trumpet and 5 guitar.

2. *Range output* methods: same as above but their output is a float value that indicates a musical instrument according to the range area it belongs to. The ranges defined in the paper under examination are 1-10 for piano, 11-20 for violin, 21-30 for flute, 31-40 for trumpet and 41-50 for guitar.

The presented system was implemented in GPLab [40] and the presented results were very encouraging. The dataset comprised 3006 audio samples from 5 musical instrument recording taken from 2 collections. The validation of results was performed in a 10–fold cross validation fashion, i.e. ten sets of 300 pieces were created, each of which preserved the analogy among all instruments. For each of these ten sets, a training–testing process was realized: testing with the considered set and training the GP system with remaining nine. The 10–fold cross validation simulations were executed for 30 different population initializations, to obtain statistics about the overall expected accuracy of the methodology. Each excerpt was initially described by 95 features, but the 14 most prominent features were finally retained by applying principal component analysis (PCA) [20]. In the work under discussion, classification accuracy percentages are provided both for the full and the PCA-reduced features set, as well as for the accuracy achieved during both the training and the testing phase. Regarding the training phase, the accuracy percentage of around 95% for the best of the 30 runs, demonstrates that the adaptation of GP was immense, since it was able to accurately "interpolate" through 2706 points in a 95–dimensional space – using the single output fitness method. Utilizing all 95 features yielded among all 30 simulations a mean test set classification accuracy in the range of 62%. By retaining the 14 most prominent features, the mean accuracy was improved to approximately 76%, a fact that amplifies the importance of selecting proper features for GP-based classification.

2.2. Sound Synthesis

Among the first works that examined automated sound synthesis through systems evolved with GP was the one presented in [35]. Therein, the author reports on some initial results yielded by a system that evolves simple mathematical functions through GP, with interactive fitness evaluation (i.e. evaluation provided by the user). These functions were producing values that constructed a waveform. Main aim of the system was to lead generations of such functions to ones that produced more pleasant sonic waveforms, since the evaluations incorporated the pleasantness itself, as perceived by the users of the system. However, as the author of the report [35] notices, not all the noises that were produced by the system sounded either pleasant or interesting.

Methodological aspects of the first sound generating GP system discussed hitherto [35], included the utilization of a typical evolutionary scheme that evolved populations using simple function combinations. Such functions included, `sin`, `cos`, `random`, `exp`, `log`, `max` and `min`, while terminals nodes included float numbers in $[-1,1]$. Through this first basic modeling the results were reported irritatingly poor to the user, resulting to noises that were hardly evolvable. An improvement was reported by dividing these functions into *amplitude* and *frequency* ones, allowing only multiplicative combinations for one function from each of these categories, i.e. functions of the form:

$$\texttt{signal}(i) = \texttt{amp}(\frac{i}{sr})\, 2\pi\, \texttt{freq}(\frac{i}{sr}), \tag{1}$$

where sr is the sample rate. Several other improvements are reported in the technical report under discussion, which are technical and do not concern the utilization of GP per se and thus are hereby omitted.

Although the methodology described in the previously discussed report did not yield the expected results, it is the first document that reveals some of the principles for evolution of sounds up to date. Two of the most important, among others, are:

1. Sound modeling plays an essential role in GP design. Thus, the power of GP is *limited* to the extent that the modeling methodology demarcates, i.e. very meticulous selection of non–terminal nodes is required.

2. In interactive evolution, it is important that the first few generations encompass interesting and non–disturbing individuals, since if the users lose focus in the first generations, then they are hardly able to drive evolution to a specific area of desired solutions.

The phenomenon discussed in point 2 above is widely known as *user fatigue* and is a matter of serious concern for all interactive evolutionary systems, not only the ones that produce artistic material. However, when evolution concerns artistic material there can hardly be a subjective target. This fact can intuitively be described as a process of *wandering optimizer*, since an indecisive user would force the algorithm to wander the space of possible artistic creations without preference to certain regions – further amplifying the fatigue on the user's side.

2.2.1. Block–based synthesis targeted to specific sounds

A different philosophy of GP algorithms for sound synthesis concerns the generation of sounds towards matching a targeted sound. Therefore, the evaluation of the produced output is not the subjective choices of users, but the objective similarity between a produced and the target sound. The distance that defines this similarity is computed through comparing some audio features extracted from both signals (generated and target). A general overview for such systems has been given in [12] where the sound generation module includes some advanced signal processing units, instead of simple functions. These processing units can either be used as terminal nodes (e.g. sound generating oscillators) or as non–terminal nodes (e.g. with additive or frequency modulation or by combining/splitting signals). Additionally, this primitive work that expanded in subsequent publications discussed in the next paragraphs, introduces an interesting fitness evaluation scheme that concerns the distance computation between the produced and the target sound not only by comparing their features, but also by having human judgment on the similarity. This aspect is interesting, since the similarity in audio features does not necessarily impose perceptual similarity, and vice versa. However, the paper under discussion does not provide any experimental evaluation on neither fitness estimation approach.

The approach discussed in the previous paragraph was materialized in two publications, [13] and [11], that focused on the generation of sounds according to targeted sounds provided by an acoustic instrument. The sound generating modules that were combined with GP included units categorized in three classes according to the input and output they were receiving and sending:

1. `Source`: 1 output (e.g. a float number indicating frequency or amplitude).

2. `Render`: 1 input (e.g. the block where an audio output device will be plugged).

3. `Type_A`: 2 inputs and 1 output (e.g. signal addition and multiplication).

4. `Type_B`: 1 input and 2 outputs (e.g. an oscillator or a filter).

Such blocks can be combined with GP and produce sounds with a great variety of characteristics. A scheme of this process is depicted in Figure 1. Since there is no human evaluation evolved, there are also no requirements for constraints during the initial steps of evolution. Therefore this approach takes advantage of the full potential of evolutionary processes, which lays in combining great numbers of randomly combined and altered individuals. The work presented in [11] describes a prototype system that implements the above described methodology, named automatic generation of sound synthesizers (AGeSS). However, there are no indications about the accuracy of the produced results, except from images of spectrograms for target and produced sounds.

Another approach that was inspired by the previously described work was presented in [7]. Therein, a similar formalization of sound producing–altering blocks was followed, but under a more sophisticated evolutionary scheme. These blocks can be divided in three categories, according to the number of children they are allowed to incorporate in the tree representation of the programs. A short description of these three block categories follows, while the reader is referred to the table in page 7 of [7] for more details.

1. No children: These blocks receive numerical parameters that define their characteristics (e.g. basic sinusoidal oscillator or LFO filters).

2. One children: These blocks receive input from two other blocks and some numerical parameter values, producing output that is an altered version of the received input (e.g. through frequency modulation or ADSR envelope application).

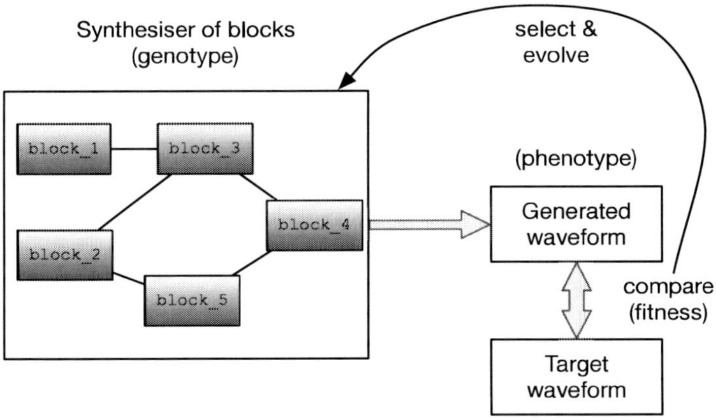

Figure 1. General design concept of evolvable sound blocks architecture for constructing sound synthesizers that reproduce a target sound.

3. Two or three children: These blocks do not include numeric parameters but receive waveform input from two signals (e.g. addition of two signals) or three signals (e.g. switch between three signals) and return a single signal.

These blocks and their parameters are evolved towards constructing synthesizers that "mimic" target sounds. The currently described thesis provides some very interesting insights about what audio similarity may concern regarding human perception; however for the purposes of the book at hand the focus remains on the evolutionary aspects of the work under discussion.

A first comment has to be made about the fact that the optimization process includes not only one but multiple samples. Therefore, the sound generators constructed through this process potentially provide a richer spectrum of audio characteristics. Additionally, there is an innovative approach regarding evolution: the introduction of a GA *"sub–optimization"* loop, that lays within the main GP evolution loop. The author of [7] refers to this sub–optimization loop as Lamarckian sub–optimization and a compact illustrative description of the proposed concept is given in Figure 2. The main GP loop is responsible for evolving the topology of the synthesizer, i.e. which blocks connect with which other, and the sub–optimization loop receives any topology given by the GP loop and optimizes its parameters by utilizing a GA, exploring the potential of every topology produced by GP. Therefore, the fitness of each individual is actually computed within the sub–optimization loop. The fitness evaluation is based on the several audio metrics that concern magnitude, phase and spectral characteristics of the produced sound, according to the audio samples that are given as target paradigms. The presented results included subjective evaluations of eight participants, who were called to rate between 0 and 10 the similarity of the target samples and the output produced by several setups of the examined algorithm. The results indicated that some evolutionary setups are better than others, while some of them outperform the previously examined methodology presented in [11].

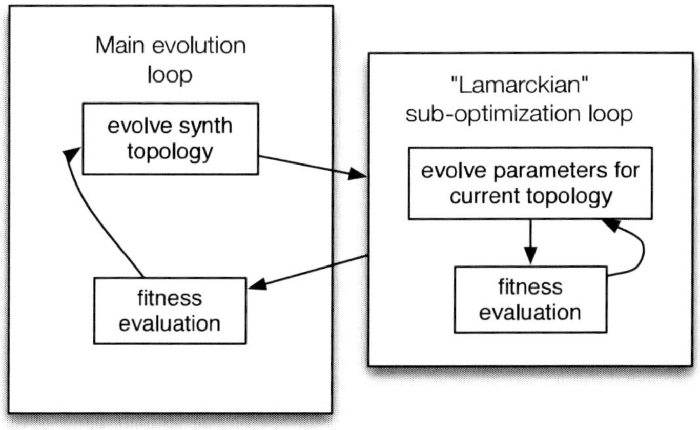

Figure 2. A compact illustrative description of the "Lamarckian" evolutionary sub–loop as described in [7].

2.2.2. Evolving physical modeling parameters

A considerable amount of work in the field of music acoustics has focused on the formulation of physical models for acoustic musical instruments [41]. On the one hand, such an approach allows for further examination and understanding of the underlying dynamics that describe the vibrating parts of an instrument. On the other hand, the model's output potentially leads to the implementation of hardware and/or software sound generators, which produce sounds that are similar to the ones reproduced by the modeled instrument. The question addressed in the paper [48] is the following: given a recorded note of a musical instrument, can the parameters of its physical model be adjusted so that the model produces a similar note event? The research in [48] examined the case of the physical model of a violin, which descriptively encompasses some basic parameters that describe not only the frequency of the reproduced note, but also expressional characteristics like the bow velocity or bow force.

To approach the problem of physical modeling, a deep understanding of signal processing is required and therefore a detailed explanation of processes in the work under discussion is beyond the scope of this book chapter. A general description of the methodology concerns the construction of functions by combining simple mathematical expressions with GP. These functions receive audio features as input and produce output that is fed to the parameters of the violin physical model. Therefore, the fitness evaluation is based on the differences between the model output and the target concerning magnitude frequency spectrum among others. The evaluation of the model was based on three different target samples: two recorded violin notes from Bach's solo partitas for violin and a third note recorded by the physical model itself. The presented results are depictions of target and produced sound spectrograms, while the author comment that the system's performance was not satisfactory. The author also proposes that a possible step towards improving the model's performance is to embody an additional model for room reverberation, since the frequencies introduced by room acoustics are not considered by the violin physical model.

3. Music Synthesis

Automated music composition is a special field of study, since a great portion of the evaluating process also address the factor of subjectivity in what sounds pleasant, or even musical. Systems that incorporate evolutionary mechanisms, as GP-based methodologies, are called to perform fitness evaluation either by interactive means, requiring from the user to provide fitness evaluation through rating simulations, or by developing fitness functions that effectively quantify qualitative characteristics of music. Both approaches have strong and weak points. Interactive evolution "models" the human preference directly, by letting the user decide for the pleasantness of the generated output. However, the implications of user fatigue do not allow a great number of evaluations, imposing restrictions to the number of individuals and generations of the evolutionary scheme, in addition to the potential misleading ratings from the user within the rating simulations – since user fatigue also affects the judgment of the user. The development of qualitative fitness functions solves the problems addressed in interactive evolution, however such functions are doubtfully effective since human subjectivity and creativity are yet imperfectly computationally modeled. The examined GP approaches are therefore presented in two categories. In the first category, advanced machine learning techniques are utilized to "learn" the aesthetic value of music pieces by themselves – thus considering no ground–truth aesthetic measures. The second category discusses interactive GP approaches. The discussion on works that examine aesthetic fitness criteria for evolutionary systems is addressed in the next section of this chapter.

3.1. Automated Critics without Aesthetic Background

The first documented approach to automated music composition using GP was reported in [43]. Music generation was based on already existing pieces, which are called "case bases". The core concept of this approach was to combine short "atom" programs (or music functions) that receive a melodic part comprising some note events, then alter these events and return as output the resulting melodic part. For example, the program named as `REP` takes a melody and returns another one that consists of four replications of the first melodies first bar. Another example is the `INVERT` program, which receives a melody and returns again a new one, which has the inverted intervals of the first. There are 13 such musical programs–functions that are utilized in the work under discussion. The altered melodies were evolved with fitness values provided by musical criteria inspired by the work presented in [4], which descriptively involve five indications of music and rhythmic novelty and balance. The results of this pioneering GP approach were produced by applying the system on a solo part of Charlie Parker. The authors commented on the results that although the fitness evaluation was found to be a good choice, the resulting music did not actually sound pleasant, a fact that indicates the importance and difficulty of describing aesthetic criteria for music.

The immense difficulty to extract music features that encompass aesthetically meaningful information is a thorough matter of discussion that is partly extended in Section 4.. The first approach in GP methodologies to tackle the problem of aesthetic definitions in music was performed by avoiding such definitions, through utilizing an artificial neural network

(ANN) as an automatic fitness rater, an approach that has proven very popular to some GP techniques that followed. The first work to introduce ANN automatic fitness raters was presented in [42], which continues the work discussed in the previous paragraph and also utilizes the GP programs–functions that were previously discussed. Therein, only 4 bar melodies are considered, with each bar including 48 equally spaced positions of potential note placements, i.e. each melody comprised 192 possible note values. The integer value of each of the 192 cells indicated the pitch height of the note, while the zero value indicating no note event. These 192 values were given as input to the ANN, which had two outputs, one indicating whether the presented melody is pleasant and one for the contrary case.

The ANN was trained using four sets of melodies, the first of which included the pleasant paradigms (Charlie Parker solos) and the remaining three included randomly altered versions of the pieces in the first category (constituting the non–pleasant paradigms for the ANN). The trained ANN exhibited the ability to recognize some sense of "musicality" since it judged as good music some Charlie Parker solos (not all of the presented ones) and also a solo by Jimi Hendrix, while it gave negative scores to some almost random sequences. The GP individuals were evolved according to the fitness indication provided by the output of the trained ANN, according to their pleasantness. The reported results were again poor generating "bizarre" rhythmic groupings, as described by the authors, and long interval jumps. According to the authors, this is primarily due to the small number of training samples for the ANN. The results were improved slightly by introducing additional explicit rules in the fitness evaluation scheme, which did penalized melodies which had a good ANN fitness but some obviously unpleasing characteristics. The interested reader is also referred to [44] for more elaborative comments of the systems described in [43] and [42].

Another approach to fitness evaluation through an "artificial expert" has been proposed in [34], where instead of an ANN, a self organizing map (SOM) [25] is utilized as an automatic fitness rater. The GP framework is similar to the ones described hitherto in this section, where simple musical programs act as non–terminal nodes (e.g. which modulate the pitch of all notes in an input melody segment or alter its durations of notes). A typical evolutionary scheme is utilized, where monophonic melodies are represented in MIDI format. The melodies are transformed to pitch class and duration vectors and are imported to a trained SOM. The SOM is trained on a target melody, capturing some of its basic pitch and duration characteristics. Thereafter, each GP individual is mapped onto the trained SOM, thus providing fitness evaluations. The presented results concern evaluations with SOMs trained on small size melodies, while the statistical comments and the small score segments that are provided do not allow a deep understanding of the system's capabilities. For further study on artificial critics without aesthetic background, the interested reader is also referred to [36] and [33].

3.2. Interactive Evaluation

Interactive evolution with GP mechanisms for generating music sequences was initially reported in [18] and [19]. The evolutionary scheme was thereby guided by the selection and ratings of a human user. However, aware of the fact that user fatigue is imminent in such systems, the authors have proposed a parallel process of non–interactive evolution, which aims to approach the objective rating behavior of the user. Before a more detailed description of the evaluation scheme, the discussed approach's formalization of music and of musical functions that are evolved by GP are introduced. The terminal nodes of the GP scheme are note pitch values within a one–octave range, and a set of seven chords. Chords are considered to be composed of three simultaneous notes, allowing the system to produce polyphonic events. Since the interpretation of notes is sequential, any note sequence without chords would be monophonic, thus reducing the interestingness of the produced melodies. The functions that compose the non–terminal nodes of the GP formalization are simple musical functions that receive as input a single note or a sequence of notes and provide a modified sequence as output. An example of a function that receives two note sequences is the `play_two` function, which simply concatenates serially the two input sequences. An example of a function with arity one is the `shift_up` function, which receives a note sequence and returns the same sequence with every pitch shifted up the next pitch value.

Fitness evaluation is given directly by the user through a graphical user interface as an integer in the scale of 1 and 100. After several simulations of user ratings, all the pairs of melodies and respective ratings were gathered to comprise a dataset for training the automatic music raters. The automatic raters incorporated the utilization of ANNs with topology and training routine that differs slightly from the standard feedforward architecture and back propagation of errors. The ANNs were accepting as input each note or chord of the melody and their output was also an integer value between 1 and 100. Thereafter, the trained ANNs were used as a substitute for human evaluation. However, the musical results obtained by the ANN runs did not sound as good as the ones produced during the human evaluation tests.

A hybrid GA and GP evolutionary strategy was presented in [47], where drum rhythms were evolved interactively. Again, an additional "user–assistance" step was also employed, in which an ANN was trained during the user rating simulations. Drum rhythms included three elements, kick, snare and hi–hat, while rhythms where genetically encoded through a rhythm matrix with three lines (one for each element) and an amount of columns equal to the length of the rhythm in sixteenths. Each integer matrix entry described the intensity of the rhythm element at the row and the given time instance at the column under discussion. The GA evolution phase incorporated the exchange and modification of genetic information in the rhythm matrices. The GP phase simply changed the arrangement of GA individuals over time, producing longer rhythmic sequences. During the user rating rounds, the matrix (genotype) representation of rhythms were given directly as input to an ANN, while the training output value was the value that the user was providing as a fitness value. Therefore, during the user runs an ANN was trained in parallel. The ANN was used to confront the user fatigue phenomenon by rejecting some rhythms that were "guaranteed" to be less interesting for the user. However the experimental results of this interesting approach did not clarify whether this objective was indeed accomplished.

An approach that incorporated not only interactive evolution but also allowed for further user intervention was presented in [1] and [2]. In these works, melodies were represented as trees, with terminal nodes being notes and non–terminal nodes being note "connectors" either serially (`S` node merges two note events one after another) or vertically (`U` node merges two note events to be played simultaneously) or other functions that rearrange or modulate note sequences. After random initialization of the population, the evolutionary process progresses by receiving user ratings for each individual–melody. Additional control is provided to the user through the user interface that accommodated the interaction between the user and the system. The user interface was providing a visual interpretation of each individual's tree representation, allowing the user to straightforwardly intervene to the genotype of each melody. Furthermore, the user was able to store some individuals in memory, removing them from the generation pool. The stored in memory individuals could be re–injected by the user into the generation pool at any time, allowing the user to re–introduce some musical characteristics that might had potentially been lost throughout the evolutionary process. A final user intervention aspect concerns the ability that was given to the users to import custom functions (non–terminal nodes) by writing a program that performs the desired alteration to musical sequences.

There are some further technical details that the previously described system encompasses regarding musical aspects and how these are expressed for being compatible with GP. However, further focus on these aspects would require extensive text to be completely described. Since the user is oriented to people that have some knowledge of music, the experimental results were produced by calling classical music composers to fill out a questionnaire in free style, i.e. to write any comments about their experience. The favorable points concerned the ability that is given to the user to refer and revise musical structure directly. On the one hand, the fact that the tree representation of melodies is easily interpreted provided the facility to generate variations of good individuals. On the other hand, a positive impact was concluded by the fact that the user was able to re–inject good characteristics stored in individuals that are stored in memory. A non–favorable comment that was made by some users regards the fact that programming skills were required to produce a custom made non–terminal node. The authors of the discussed papers conclude that the utilization of a visual programming language could improve the custom function insertion process.

The approaches that have been discussed hitherto in the current section, consider the application of GP–generated functions on notes from some initial pieces which were either given by an existing musical piece ([43], [42]), or produced randomly ([47], [1], [2]). The proposal to utilize a rule of physics to create an initial melody for feeding a GP system was proposed in [3]. Specifically, the rule of the "rubber band" to create an initial jazz solo that complies with some initial principles, while afterwards, the application of simple functions evolved with GP alter the initial solos, generating novel jazz solos that constitute variations of the initial solo. The authors of [3] backup their approach by providing an interesting comment that concerns jazz improvisation:

> *A good improviser is not expected to drop the focus and give way to meaningless cascades of notes or producing routine phrases for lack of artistic ideas. The challenge is to be able to plan the structure of the solo already from start, and then stick to the plan during the entire solo.*

Based on this comment, they decided upon applying the rubber band rule to plan an initial solo structure. The initial melodic line of the solo is constructed by following a "rubber band" principle, according to which each pitch interval is selected by computing some energy constraints simulating the tension of a stretched rubber band. Larger intervals are thereby assumed to require more energy, while the energy is also dependent on the length and volume of the destination note. An initial solo is generated by considering a fixed arbitrary amount of energy to be consumed for the rubber band that describes it, a fact that potentially allows some areas of solos to be include large intervals, forcing others to include small ones. After the initial construction, the solo is divided in a a predefined number segments, called the "*delta phrases*". The delta phrases are then submitted to alterations based on some simple programs–functions. Examples of such functions include transposition of the entire delta phrase, random rhythm alteration etc. The functions are combined in a tree–like structure. Different trees, representing different alterations of the initial solo, are evolved using GP, while the fitness is provided by user ratings. The presented results include short descriptions of certain solos and providing web links for listening to some of them. The authors of [3] concluded that they should also examine the efficiency of the system with an automatic (non–interactive) fitness evaluation process. For further study on interactive evolution the reader is referred to [39], where the interactive process takes the form of a virtual instrument through a Nintendo Wii controller, as well as to [45].

A work that differentiates from all the interactive GP approaches described above is presented in [28]. Again, this work incorporates the subjective evaluation of polyphonic melodies that are evolved by GP, however the individuals–melodies that are constructed through the subjective evaluations are stored on an online database, giving the opportunity to many users to provide their ratings, contributing to the evolution of novel melodies. By the time that [28] had been written, the population was evolved for 2513 generations, with the participation of 6931 volunteers. Aim of the experiment was to examine the role that "consumer" selection plays to music generation, apart from several other social and geographical accounts that might be attributed to the evolution of music. The initially random loops where very quickly evolved to melodies that encompass musical attributes, like pleasing chord transitions and rhythm. However, after the first "fitness explosion", the improvement of individuals was decelerated leading to a steady plateau of mean ratings significantly below the perfect rate.

4. Aesthetic Criteria for Fitness Evaluation

In several of the works presented in Section 3.1., the evolutionary process was led by fitness evaluations provided by "artificial critics". The critics were mostly statistical systems (like ANNs or SOMs) that received raw music data as input, i.e. the explicit note and duration values, and were trained to judge whether the input data corresponded to "good" or "bad" music. A different approach was followed in [33], where the role of the critic was substituted by a process that included feature extraction from the generated music. Specifically, 24 features were extracted from the melodies produced by the generative system presented therein, which comprised the 24–dimensional feature vectors that represented each piece. These features included general information in the form of statistics about note and rhythm distributions. The fitness evaluation was performed by computing the distance between the feature vector of a piece and a "benchmark" feature vector. This benchmark feature vector however, was conventionally chosen by a process that does not include any information about wether this vector describes a pleasant melody. The target feature vector was selected to include the median value respective feature in each dimension, a process that simply allowed the authors in [33] to reach conclusions about the fact that the proposed GP system was in fact converging to the targeted feature area. This section discusses the research trend towards harnessing evolutionary music composition techniques, and especially GP techniques, with a parallel examination of sophisticated music features.

Among the first approaches to extracting music features that encompass aesthetic meaning, discussed the extraction of features through the application of the Zipf–Mandelbrot law on musical pieces encoded in MIDI. A thorough description of these features is beyond the scope of this book chapter and therefore a brief overview of the aforementioned features is given. The work presented in [30] was based in the hypothesis that the application of the Zipf–Mandelbrot law would allow a computational identification of aesthetic aspects of music. Specifically, an initial set of attributes (metrics) of music pieces was detected and the Zipf–Mandelbrot law was applied on the resulting distributions. These metrics included several musical aspects like pitch and duration of musical events, combinations of pitch and duration of musical events among several others. The results presented therein, and in subsequent studies [29,31] indicated that metrics deriving from the application of the Zipf–Mandelbrot on statistical distributions of several metrics could very effectively categorize music from a wide variety of genres (baroque, classical, 20th century, blues, jazz, etc.) This study has also led to the implementation of a music recommendation system based on these metrics, as extracted from mp3 files in vast music libraries [17].

Based on the metrics discussed in the previous paragraph, a GP based approach was also employed in [32] for music composition. Initially, the features of 992 musical pieces from various genres were utilized to train ANNs, yielding a percentage of 87.85% accuracy on classification according to the popularity of each piece. Afterwards, following the assumption that popularity correlates with aesthetic value, these ANNs were employed as fitness raters to an into a GP music composition system named NEvMuse. This system was utilized to autonomously (without human evaluation) compose novel variations of a Bach's Invention (BWV 784). These variations were not only found aesthetically pleasing, but also the aesthetic judgment of 23 human evaluators significantly correlated with the fitness evaluations provided by the ANN critic.

The remaining of this section focuses on two approaches that utilize GP to evolve functions that pertain to a special class of functions, which are able to render waveforms that present interesting structure in many levels of sound (from *micro* to *meso*, as described in [37]), producing sounds an noises that aesthetically refer to early pulse code modulation (PCM) 8–bit melodies. Therefore, when addressing the output of these functions, the terms "sounds" and "melodies" will be used interchangeably. These functions are described in [15] and instantly attracted the attention of the 8–bit music community [14]. Terminal nodes of these functions are integers, while non–terminal nodes include not only typical arithmetic operators ("+", "-", "*", "/") but also bitwise operators like `AND` (`&`), `OR` (`|`), `XOR` (`^`), left bit shift («) and right bit shift (»). Consequently, the application of GP for evolving functions that are composed of the aforementioned setup of terminal and non–terminal nodes is quite straightforward, considering their tree representation. The evolution of these functions through interactive GP has been utilized for exploring the audio feature space towards finding aesthetic measures for sound [22] and for developing fitness adaptive operators to reduce the phenomenon of user fatigue [21].

4.1. Genetic Programing, Aesthetics and User Fatigue

The work presented in [22] utilized an interactive GP evolutionary scheme that allowed users to rate melodies produced by the previously described functions through a graphical user interface, providing fitness evaluation for evolution. During the evolution simulations, three categories of features were documented for each melody: waveform features (describing properties of the waveforms), spectral features (describing features about their spectrum), as well as several features describing the structure that the tree representation of the functions that produced these melodies. An example of a waveform feature was the Shannon information entropu [38] of a quantized version of the waveform. Among the spectral features the computation of the Mel–Frequency Cepstral Coefficients (MFCCs) [6] was included, since these features have proven useful for measuring the perceptual similarity of audio samples [46]. Within the tree structure features was the operator entropy (providing indications about the distributions of operators in individuals.)

Through subjective tests over 9 participants, it was initially evaluated that the system was indeed producing melodies that were more highly rated as the generations progressed. Therefore, the total improvement of rating between the initial and the final generation indicated that the melodies in the final generation encompassed higher aesthetic values for the users. It has to be noted that the improvement in ratings was not measured directly by the rating values, since on the one hand different users have different rating profiles (e.g. a rate of 5 out of 10 might indicate a good or a bad individual according to the strictness in rating) and on the other hand, the ratings depend on the initialization: a good initial population might have less chances to improve impressively. Therefore, the improvement in ratings was measured through the *relative differences* between the ratings in the final and the initial generations, i.e. the ratio of the difference between final and initial generation ratings over the initial generation ratings. Except from the ratings, some of the extracted features were also found to be significantly different from the initial population to the final generation. Thereby, in [22] it was discussed whether this process could indicate which features are "responsible" for recognizing subjective or objective aesthetically pleasing melodies. It was also proposed that evolutionary systems might be a perfect methodology for exploring such questions, since the produced melodies–individuals include a wide spectrum of possibilities in the interestingness of the output: from completely random and "false" melodies to completely structured and "boring" ones.

A modification of the system described previously was presented in [21], which introduced the fitness–adaptive operators for GP. These operators were indicated to partially alleviate the phenomenon of user fatigue by allowing the users to have a sense of "control" in the expected randomness in the individuals of the forthcoming generation. For supporting the motivation of this work, an initial examination was conducted by measuring the phenotypical distance between parent and children nodes, after applying genetic operators that were acting on different heights of the genotypical (tree) representation. Specifically, two experimental processes were followed that statistically indicated the following relation: the closer to the root an operator acted, the more differences was observed between the phenotypes of parents and children. These differences were reflected in the waveform and spectral features discussed previously. It should be noted however, that the work under discussion did not aim to generalize this hight–fitness relationship to a wider GP context; it was just observed and exploited in the context of the presented system.

Aim of the work in [21] was to make the 8–bit GP system more user friendly, by allowing users to control the randomness of the individuals in the subsequent generations. The evolution of individuals in interactive evolutionary systems progresses rather slowly for human evaluators, since often a considerable amount of individuals have to be rated in order to reach individuals with good fitness. While computers have no problem to iteratively "rate" vast numbers of individuals in several generations, human users sense the impediment of fatigue within the first few dozens of ratings. The evolutionary process is further delayed by the fact that highly rated individuals might produce offspring that are completely different and most possibly less fit. Considering the comments in the previous paragraph, according to which the higher the depth that a GP operator acts the small the difference of the produced individuals, the "amount" of change may be defined by the fitness – rate of an individual. Specifically, for computing the depth of operator action (d_i) for the individual i, the following formula has been proposed:

$$d_i = \lfloor \beta_i + 0.5 \rfloor, \tag{2}$$

where

$$\beta_i = m + \frac{f_i \cdot (D_i - m)}{F}, \tag{3}$$

where m is the lowest considered depth of operator action, D_i is the total depth of the i individual's tree representation, f_i is its fitness value (rate), with lower considered fitness with a value 0 and higher F (regarding the current rating simulation). However, the depth d_i does not provide a sense of control of the randomness in the evolutionary process. This was achieved by introducing a *risk factor parameter* (RF) which defined the "probability" that the depth of operator action proposed by 2 would be selected. Specifically, before each generation the user selected a value for RF $\in [0,1]$ – with higher values indicating higher risk not to respect the depth limitations of 2, and the final depth was selected by random selection from a normal distribution defined by

$$p(x;\beta_i,\sigma(\mathrm{RF})^2) = \frac{1}{\sqrt{2\pi\sigma(\mathrm{RF})^2}} e^{-\frac{(x-\beta_i)^2}{2\sigma(\mathrm{RF})^2}}. \tag{4}$$

The reported results among 11 participants indicated that the facilitation introduced by the RF parameter not only made the process less fatiguing, but also improved significantly the overall ratings of individuals. Specifically, the number of generations evolved by the users in the RF variation of the system was significantly larger that the random depth version, with mean values around 10 and around 7 respectively. Additionally, the improvement in ratings, measured as the relative improvement as discussed earlier, was also significantly different. Both random depth and RF variations provided individual that were better rated by the users, with a significantly higher improvement for the RF variation, indicating the positive impact of the introduced RF parameter.

5. Conclusion

In this chapter the utilization of genetic programming (GP) for several musical tasks has been studied, by providing an overview of several methodologies that have been proposed from 1994 to the date that these lines were being written. The promising power that GP has introduced to the field of computational intelligence, gave rise to methodologies that produce sound and music. The automated production of sounds aimed primarily to construct sound generators that are able to accurately reproduce a target sound (e.g. from an acoustic musical instrument). Additionally, the utilization of GP has yielded important results towards the constructing systems that perform musical instrument recognition, i.e. identify the music instrument that is incorporated in a recorded audio segment. Automatic music generation aimed to construct systems that produced pleasant music without the compositional intervention of a human expert. The idea behind such systems was to utilize evolution to either combine short programs that produce/analyze sound or music, or rearrange segments of mathematical expressions to produce waveforms or notes that are better fit to a specific task. The problem that was very soon addressed, however, was the formulation of the fitness itself, since interactive evolutionary systems were limited to small numbers of ratings by the users, while for the non–interactive systems it was difficult to devise automatic "recognizers" of musical quality.

For instrument identification, the methodological context that prevailed incorporated the extraction of sound features from recorded audio segments and the utilization of GP to construct regression models that used these features as input, providing output that indicated the existence of a specific instrument in the recorded segment. The results were very promising, indicating that identification accuracies above 90% could be reached. For audio synthesis the proposed methodologies focused either on interactive evolution of GP programs that provided the user with pleasing sounds, or the non–interactive generation of sounds that resemble the sound produced by acoustic instruments. The interactive approaches were not so efficient, since the user fatigue was very intense, further amplified by the fact that several sounds–individuals were noisy. Non–interactive evolution has given good results at some extent, as also indicated by subjective tests of target–to–generated sound comparison.

For automated symbolic music composition, the main concern was to combine programs that received small melodic segments, modulated and merged them providing integrated melodic parts. The primary aspects of fitness computation incorporated automated evaluation of the produced music according to simple musical rules. Soon however, it was realized that musical quality can hardly be described by rules, leading to the utilization of machine learning techniques (e.g. artificial neural networks (ANNs) or self organizing maps (SOMs)). Machine learning tolls like ANNs and SOMs were trained with the phenotype (notes and rhythm) of human composed melodies (like Charlie Parker solos), expectedly "internalizing" some qualitative principles within their knowledge–encoding mechanisms. However, the results were again not so impressive as expected. Afterwards, there was a great shift towards the exploration of music features that might incorporate music quality. Such features have yielded interesting results regarding music categorization, as well as popular song recognition, a fact that indicated that these features may describe qualitative musical meaning. Thereby, their utilization as fitness "targets" has proven a promising technique. As for the future prospects in GP (and evolutionary in general) music composition, the exploration of such features and of mechanisms to reduce user fatigue fatigue seem to provide a fertile research field.

References

[1] Ando, D., Iba, H.: Interactive composition aid system by means of tree representation of musical phrase. In: IEEE Congress on Evolutionary Computation, 2007. CEC 2007. pp. 4258–4265 (Sep 2007)

[2] Ando, D., Dahlsted, P., Nordahl, M.G., Iba, H.: Interactive gp with tree representation of classical music pieces. In: Proceedings of the 2007 EvoWorkshops 2007 on EvoCoMnet, EvoFIN, EvoIASP,EvoINTERACTION, EvoMUSART, EvoSTOC and EvoTransLog: Applications of Evolutionary Computing. pp. 577–584. Springer-Verlag, Berlin, Heidelberg (2007)

[3] Bäckman, K., Dahlstedt, P.: A generative representation for the evolution of jazz solos. In: Proceedings of the 2008 Conference on Applications of Evolutionary Computing. pp. 371–380. Evo'08, Springer-Verlag, Berlin, Heidelberg (2008)

[4] Baker, D.: David Baker's Jazz Improvisation: A Comprehensive Method for All Musicians. Alfred Music Publishing (Jan 1988)

[5] Brown, J.C.: Computer identification of musical instruments using pattern recognition with cepstral coefficients as features. *The Journal of the Acoustical Society of America* 105(3), 1933–1941 (Mar 1999)

[6] Davis, Mermelstein, P.: Comparison of parametric representations for monosyllabic word recognition in continuously spoken sentences. IEEE Transactions on Acoustics Speech and Signal Processing 28(4), 357–366 (1980)

[7] Donahue, C.: Applications of genetic programming to digital audio synthesis. Ph.D. thesis, The University of Texas at Austin, Department of Computer Science, Texas, USA (Dec 2013)

[8] Eronen, A.: Comparison of features for musical instrument recognition. In: IEEE Workshop on the, Applications of Signal Processing to Audio and Acoustics, 2001. pp. 19–22. New York, NY (2001)

[9] Eronen, A., Klapuri, A.: Musical instrument recognition using cepstral coefficients and temporal features. In: 2000 IEEE International Conference on Acoustics, Speech, and Signal Processing, 2000. ICASSP '00. Proceedings. vol. 2, pp. II753–II756 vol.2 (2000)

[10] Essid, S., Richard, G., David, B.: Musical instrument recognition by pairwise classification strategies. IEEE Transactions on Audio, Speech, and Language Processing 14(4), 1401–1412 (Jul 2006)

[11] Garcia, R.A.: Automating the design of sound syntheses techniques using evolutionary methods. In: Proceedings of the COST G-6 Conference on Digital Audio Effects (DAFX-01),. Limerick, Ireland (December 2001)

[12] Garcia, R.A.: Towards the automatic generation of sound synthesis techniques: Preparatory steps. In: AES 109th Convention. Los Angeles (22-25 September 2000)

[13] Garcia, R.A.: Growing sound synthesizers using evolutionary methods. Synthesis M(Garcia 2000), 99–107 (2001)

[14] Heikkilä, V.: countercomplex–bitwise creations in a pre–apocalyptic world. http://countercomplex.blogspot.com/ (Dec 2011)

[15] Heikkilä, V.: Discovering novel computer music techniques by exploring the space of short computer programs. CoRR abs/1112.1368 (Dec 2011)

[16] Herrera-Boyer, P., Amatriain, X., Batlle, E., Serra, X.: Towards instrument segmentation for music content description: a critical review of instrument classification techniques. In: ISMIR. Plymouth, Massachusetts (2000)

[17] Hughes, D., Manaris, B.: Fractal dimensions of music and automatic playlist generation: Similarity search via MP3 song uploads. In: 2012 Eighth International Conference on Intelligent Information Hiding and Multimedia Signal Processing (IIH-MSP). pp. 436–440. Piraeus, Greece (Jul 2012)

[18] Johanson, B.: Automated Fitness Raters for GP-Music System. Master's thesis, School of Computer Science, University of Birmingham, Birmingham, UK (1997)

[19] Johanson, B.E., Poli, R.: GP-Music: an interactive genetic programming system for music generation with automated fitness raters. Tech. Rep. CSRP-98-13, University of Birmingham, School of Computer Science (May 1998)

[20] Jolliffe, I.: Principal Component Analysis. Springer, 2nd edn. (Oct 2002)

[21] Kaliakatsos-Papakostas, M.A., Epitropakis, M.G., Floros, A., Vrahatis, M.N.: Controlling interactive evolution of 8-bit melodies with genetic programming. Soft Computing 16, 1997–2008 (Dec 2012)

[22] Kaliakatsos-Papakostas, M.A., Epitropakis, M.G., Floros, A., Vrahatis, M.N.: Interactive evolution of 8-bit melodies with genetic programming towards finding aesthetic measures for sound. In: Proceedings of the 1st International Conference on Evolutionary and Biologically Inspired Music, Sound, Art and Design, EvoMUSART 2012, Malaga, Spain. LNCS, vol. 7247, pp. 140–151. Springer Verlag (apr 2012)

[23] Kaminsky, I., Materka, A.: Automatic source identification of monophonic musical instrument sounds. In: IEEE International Conference on Neural Networks, 1995. Proceedings. vol. 1, pp. 189–194 vol.1. Perth, WA (Nov 1995)

[24] Kobayashi, Y.: Automatic generation of musical instrument detector by using evolutionary learning method. In: Hirata, K., Tzanetakis, G., Yoshii, K. (eds.) ISMIR. pp. 93–98. International Society for Music Information Retrieval (2009)

[25] Kohonen, T., Schroeder, M.R., Huang, T.S. (eds.): Self-Organizing Maps. Springer-Verlag New York, Inc., Secaucus, NJ, USA, 3rd edn. (2001)

[26] Loughran, R., Walker, J., O'Neill, M., McDermott, J.: Genetic programming for musical sound analysis. In: Machado, P., Romero, J., Carballal, A. (eds.) Proceedings of the 1st International Conference on Evolutionary and Biologically Inspired Music, Sound, Art and Design, EvoMUSART 2012. LNCS, vol. 7247, pp. 176–186. Springer Verlag, Malaga, Spain (11-13Apr 2012)

[27] Loughran, R.B.: Music Instrument Identification with Feature Selection Using Evolutionary Methods. Ph.D. thesis, University of Limerick, Ireland (Sep 2009)

[28] MacCallum, R.M., Mauch, M., Burt, A., Leroi, A.M.: Evolution of music by public choice. Proceedings of the National Academy of Sciences 109(30), 12081–12086 (2012)

[29] Machado, P., Romero, J., Manaris, B., Santos, A., Cardoso, A.: Power to the critics - a framework for the development of artificial critics. In: Proceedings of 3rd Workshop on Creative Systems, 18 th International Joint Conference on Artificial Intelligence (IJCAI 2003. pp. 55–64. Acapulco, Mexico (August 2003)

[30] Manaris, B., Purewal, T., McCormick, C.: Progress towards recognizing and classifying beautiful music with computers - MIDI-encoded music and the Zipf-Mandelbrot law. In: Proceedings IEEE SoutheastCon, 2002. pp. 52–57. IEEE (2002)

[31] Manaris, B., Romero, J., Machado, P., Krehbiel, D., Hirzel, T., Pharr, W., Davis, R.B.: Zipf's law, music classification, and aesthetics. Computer Music Journal 29(1), 55–69 (Apr 2005), ArticleType: research-article / Full publication date: Spring, 2005 / Copyright ÂI' 2005 The MIT Press

[32] Manaris, B., Roos, P., Machado, P., Krehbiel, D., Pellicoro, L., Romero, J.: A corpus-based hybrid approach to music analysis and composition. In: Proceedings of the 22nd national conference on Artificial intelligence - Volume 1. pp. 839–845. AAAI Press (2007)

[33] McDermott, J., O'Reilly, U.M.: An executable graph representation for evolutionary generative music. In: Proceedings of the 13th Annual Conference on Genetic and Evolutionary Computation. pp. 403–410. GECCO '11, ACM, New York, NY, USA (2011)

[34] Phon-Amnuaisuk, S., Law, E., Kuan, H.: Evolving music generation with som-fitness genetic programming. In: Giacobini, M. (ed.) Applications of Evolutionary Computing, Lecture Notes in Computer Science, vol. 4448, pp. 557–566. Springer Berlin / Heidelberg (2007)

[35] Putnam, J.B.: Genetic programming of music (30 August 1994)

[36] Reddin, J., McDermott, J., O'Neill, M.: Elevated pitch: Automated grammatical evolution of short compositions. In: Giacobini, M., De Falco, I., Ebner, M. (eds.) Applications of Evolutionary Computing, EvoWorkshops2009: EvoCOMNET, EvoENVIRONMENT, EvoFIN, EvoGAMES, EvoHOT, EvoIASP, EvoINTERACTION, EvoMUSART, EvoNUM, EvoPhD, EvoSTOC, EvoTRANSLOG. LNCS, vol. 5484, pp. 579–584. Springer Verlag, Tubingen, Germany (15-17 Apr 2009)

[37] Roads, C.: Microsound. MIT Press (2004)

[38] Shannon, C.E.: A mathematical theory of communication.*ACM SIGMOBILE Mobile Computing and Communications Review* 5, 3–55 (January 2001)

[39] Shao, J., McDermott, J., O'Neill, M., Brabazon, A.: Jive: A generative, interactive, virtual, evolutionary music system. In: Proceedings of the 2010 International Conference on Applications of Evolutionary Computation - Volume Part II. pp. 341–350. EvoCOMNET'10, Springer-Verlag, Berlin, Heidelberg (2010)

[40] Silva, S., Almeida, J.: Gplab-a genetic programming toolbox for matlab. Proceedings of the Nordic MATLAB conference pp. 273–278 (2003)

[41] Smith, J.O.: Physical audio signal processing: for virtual musical instruments and audio effects. W3K Publishing, Stanford (2010)

[42] Spector, L., Alpern, A.: Induction and recapitulation of deep musical structure. In: Proceedings of the IFCAI-95 Workshop on Artificial Intelligence and Music. pp. 41–48 (1995)

[43] Spector, L., Alpern, A.: Criticism, culture, and the automatic generation of artworks. In: Proceedings of the twelfth national conference on Artificial intelligence (vol. 1). pp. 3–8. AAAI '94, American Association for Artificial Intelligence, Menlo Park, CA, USA (1994)

[44] Spector, L., Klein, J., Harrington, K.: Selection songs: Evolutionary music computation. *YLEM Journal (Artists Using Science and Technology)* 25(6 & 8), 24–26 (spring 2005)

[45] Tanji, M., Iba, H.: Conbreo: A music performance rendering system using hybrid approach of iec and automated evolution. In: Proceedings of the 12th Annual Conference on Genetic and Evolutionary Computation. pp. 1275–1282. GECCO '10, ACM, New York, NY, USA (2010)

[46] Terasawa, H., Slaney, M., Berger, J.: Perceptual Distance in Timbre Space. In: Proceedings of the 11th International Conference on Auditory Display (ICAD2005). pp. 61–68. Department of Computer Science and Information Systems, University of Limerick, Limerick, Ireland (2005)

[47] Tokui, N.: Music composition with interactive evolutionary computation. *Communication* 17(2), 215–226 (2000)

[48] Wilson, R.S.: First steps towards violin performance extraction using genetic programming. In: Koza, J.R. (ed.) Genetic Algorithms and Genetic Programming at Stanford 2002, pp. 253–262. Stanford Bookstore, Stanford, California, 94305-3079 USA (June 2002)

INDEX

#

A

B

D

E

F

G

H

I

J

K

L

M

N

O

P

Q

R

S

T

U

V

W

X

Y

Z